Test Bank

Contemporary Business Mathematics
for Colleges
FOURTEENTH EDITION

James E. Deitz
Heald Colleges

James L. Southam
San Francisco State University

THOMSON

SOUTH-WESTERN

Australia · Brazil · Canada · Mexico · Singapore · Spain · United Kingdom · United States

THOMSON

SOUTH-WESTERN

Test Bank to accompany Contemporary Business Mathematics for Colleges, 14th Edition
James E. Deitz and James L. Southam

VP/Editorial Director:
Jack W. Calhoun

Editor-in-Chief:
Alex von Rosenberg

Senior Acquisitions Editor:
Charles McCormick

Senior Developmental Editor:
Alice Denny

Marketing Manager:
Larry Qualls

Production Project Manager:
Margaret M. Bril

Manager of Technology, Editorial:
Vicky True

Technology Project Editor:
Christine Wittmer

Manufacturing Coordinator:
Diane Lohman

Production House:
DPS Associates

Printer:
Globus Printing
Minster, OH

Art Director:
Chris A. Miller

Ancillary Coordinator:
Stephanie Schempp

Ancillary Cover Designer:
Patti Hudepohl

For permission to use material from this text or product, submit a request online at http://www.thomsonrights.com.

For more information about our products, contact us at:

Thomson Learning Academic Resource Center

1-800-423-0563

Thomson Higher Education
5191 Natorp Boulevard
Mason, OH 45040
USA

TABLE OF CONTENTS

Chapter 1   FUNDAMENTAL PROCESSES

PROBLEMS

LEARNING OBJECTIVE 1

1.
```
   7
   3
   9
   1
   8
 + 2
```

SOLUTION: 30

2.
```
  12
  18
  59
  61
  17
 + 33
```

SOLUTION: 200

3.
```
  564
  216
  455
  155
  392
 + 218
```

SOLUTION: 2,000

4.
```
  1,514
  2,211
  7,142
  5,763
 + 5,875
```

SOLUTION: 22,505

5.
  2,347
  9,247
  3,002
  7,666
 +1,020

SOLUTION: 23,282

6.
  14,307
  20,316
  88,404
  29,330
 + 92,041

SOLUTION: 244,398

7.
  1,342,716
  2,242,111
  5,076,223
  8,401,776
 + 6,632,218

SOLUTION: 23,695,004

8.
  306
  928
  788
  401
  992
 + 827

SOLUTION: 4,242

9.
$7 + 3 + 2 + 8 + 1 + 9 =$ _____

SOLUTION: 30

10.
$9 + 6 + 5 + 8 + 7 + 4 + 6 =$ _____

SOLUTION: 45

11.
$87 + 13 + 46 + 54 + 72 + 28 =$ _____

SOLUTION: 300

12.
$237 + 682 + 921 + 879 + 786 = $ _____

SOLUTION: 3,505

13.
$1,850 + 1,150 + 2,000 + 3,500 = $ _____

SOLUTION: 8,500

14.
$42,317 + 37,204 + 92,076 + 23,076 = $ _____

SOLUTION: 194,673

15.
$5 + 5 + 7 + 3 = $ _____
$3 + 4 + 6 + 7 = $ _____
$1 + 1 + 5 + 8 = $ _____
$\underline{1} + \underline{1} + \underline{1} + \underline{2} = $ _____

__ + __ + __ + __ = _____

SOLUTION:

Vertical	$10 + 11 + 19 + 20 = 60$
Horizontal	$20 + 20 + 15 + \ \ 5 = 60$

16.
$9 + 4 + 8 + 6 = $ _____
$7 + 9 + 4 + 3 = $ _____
$9 + 6 + 2 + 4 = $ _____
$\underline{8} + \underline{5} + \underline{9} + \underline{7} = $ _____

__ + __ + __ + __ = _____

SOLUTION:

Vertical	$33 + 24 + 23 + 20 = 100$
Horizontal	$27 + 23 + 21 + 29 = 100$

17.
49 + 54 + 68 + 76 = _____
57 + 49 + 34 + 23 = _____
89 + 76 + 62 + 34 = _____
<u>78</u> + <u>95</u> + <u>69</u> + <u>87</u> = _____

___ + ___ + ___ + ___ = _____

SOLUTION:

Vertical $273 + 274 + 233 + 220 = 1,000$
Horizontal $247 + 163 + 261 + 329 = 1,000$

18.
83
-52

SOLUTION: 31

19.
108
-97

SOLUTION: 11

20.
236
-192

SOLUTION: 44

21.
987
-212

SOLUTION: 775

22.
700
-359

SOLUTION: 341

23.
4,892
$-3,471$

SOLUTION: 1,421

24.
  5,564
− 2,687

SOLUTION: 2,877

25.
87 − 44 = _____

SOLUTON: 43

26.
97 − 28 = _____

SOLUTION: 69

27.
643 − 512 = _____

SOLUTION: 131

28.
457 − 163 = _____

SOLUTION: 294

29.
534 − 145 = _____

SOLUTION: 389

30.
  98 − 34 = _____
− 67 − 24 = _____

__ − __ = _____

SOLUTION:

Vertical   31 − 10 = 21
Horizontal 64 − 43 = 21

31.

$83 - 36 =$ _____

$- \underline{44} - \underline{17} =$ _____

__ - __ = _____

SOLUTION:

Vertical	$39 - 19 = 20$
Horizontal	$47 - 27 = 20$

32.

56
$\times\ 4$

SOLUTION: 224

33.

39
$\times 7$

SOLUTION: 273

34.

231
$\times\ 30$

SOLUTION: 6,930

35.

306
$\times\ 90$

SOLUTION: 27,540

36.

998
$\times\ 26$

SOLUTION: 25,948

37.

292
$\times\ 31$

SOLUTION: 9,052

38.
728
× 400

SOLUTION: 291,200

39.
601
× 301

SOLUTION: 180,901

40.
628
× 628

SOLUTION: 394,384

41.
1004
×   203

SOLUTION: 203,812

42.
$798 \times 100 = \underline{\hspace{1cm}}$

SOLUTION: 79,800

43.
$17,008 \times 25 = \underline{\hspace{1cm}}$

SOLUTION: 425,200

44.
$14,000 \times 50 = \underline{\hspace{1cm}}$

SOLUTION: 700,000

45.
$48 \div 2 = \underline{\hspace{1cm}}$

SOLUTION: 24

46.
$369 \div 3 = \underline{\hspace{1cm}}$

SOLUTION: 123

47.
$1,648 \div 4 =$ _____

SOLUTION: 412

48.
$97,630 \div 10 =$ _____

SOLUTION: 9,763

49.
$1,000,000 \div 100 =$ _____

SOLUTION: 10,000

50.
$600,000 \div 50 =$ _____

SOLUTION: 12,000

51.
$900 \div 60 =$ _____

SOLUTION: 15

52.
$17\overline{)408}$

SOLUTION: 24

53.
$12\overline{)444}$

SOLUTION: 37

54.
$14\overline{)2,436}$

SOLUTION: 174

LEARNING OBJECTIVE 5

55.
Give an estimated answer: 19×31

SOLUTION: 600

56.
Give an estimated answer: 71×89

SOLUTION: 6,300

57.
Give an estimated answer: 42×48

SOLUTION: 2,000

58.
Give an estimated answer: 297×306

SOLUTION: 90,000

59.
Give an estimated answer: $4,489 \times 299$

SOLUTION: 1,350,000

60.
Give an estimated answer: $9,999 \times 9,999$

SOLUTION: 100,000,000

Chapter 2   FRACTIONS

PROBLEMS

LEARNING OBJECTIVES 1, 2

1.
Change the improper fractions to whole or mixed numbers. Reduce fractional parts to lowest terms.

a. $\dfrac{95}{15}$ b. $\dfrac{50}{4}$ c. $\dfrac{85}{51}$ d. $\dfrac{75}{54}$

SOLUTION:
a.   6 1/3   b.   12 1/2   c.   1 2/3   d.   1 7/18

2.
Change the improper fractions to whole or mixed numbers. Reduce fractional parts to lowest terms.

a. $\dfrac{51}{6}$ b. $\dfrac{63}{9}$ c. $\dfrac{414}{15}$

SOLUTION:
a.   8 1/2   b.   7   c.   27 3/5

LEARNING OBJECTIVE 1

3.
Change the mixed numbers to improper fractions.

a. $8\dfrac{3}{5}$ b. $4\dfrac{5}{8}$ c. $4\dfrac{5}{6}$

SOLUTION:
a.   43/5   b.   37/8   c.   29/6

4.
Change the mixed numbers to improper fractions.

a. $3\dfrac{5}{6}$ b. $9\dfrac{3}{8}$ c. $2\dfrac{7}{11}$

SOLUTION:
a.   23/6   b.   75/8   c. 29/11

LEARNING OBJECTIVE 2

5.
Change each fraction to higher terms, with the denominator shown.

a. $\dfrac{4}{7} = \dfrac{}{42}$

b. $\dfrac{3}{5} = \dfrac{}{35}$

c. $\dfrac{3}{4} = \dfrac{}{52}$

SOLUTION:
a. 4/7 = 24/42

b. 3/5 = 21/35

c. 3/4 = 39/52

6.
Change each fraction to higher terms, with the denominator shown.

a. $\dfrac{5}{8} = \dfrac{}{24}$

b. $\dfrac{3}{13} = \dfrac{}{65}$

c. $\dfrac{11}{12} = \dfrac{}{72}$

SOLUTION:
a. 5/8 = 15/24

b. 3/13 = 15/65

c. 11/12 = 66/72

LEARNING OBJECTIVES 2, 3

7.
Add the following fractions and mixed numbers. Write the answers as fractions or mixed numbers in lowest terms.

a. $\dfrac{3}{4}$

$+\dfrac{5}{6}$

b. $\dfrac{1}{6}$

$\dfrac{3}{8}$

$+\dfrac{3}{5}$

c. $\dfrac{1}{2}$

$\dfrac{2}{3}$

$\dfrac{1}{4}$

$+\ \dfrac{3}{5}$

SOLUTION:

a. 3/4 = 9/12
    + 5/6 = 10/12
    19/12 = 1 7/12

b. 1/6 =   20/120
    3/8 =   45/120
    + 3/5 =   72/120
    137/120 = 1 17/120

c. 1/2 = 30/60
    2/3 = 40/60
    1/4 = 15/60
    + 3/5 = 36/60
    121/60 = 2 1/60

8.

Add the following fractions and mixed numbers. Write the answers as fractions or mixed numbers in lowest terms.

a. $2\dfrac{7}{8}$

$+1\dfrac{1}{6}$

b. $4\dfrac{3}{4}$

$2\dfrac{9}{14}$

$+\dfrac{6}{7}$

c. $\dfrac{4}{5}$

$1\dfrac{3}{4}$

$3\dfrac{5}{8}$

$+\dfrac{7}{10}$

SOLUTION:

a. 2 7/8   =   2 21/24
$\underline{+ 1\ 1/6\ \ =\ \ \ \underline{1\ \ 4/24}}$
            3 25/24 =   4 1/24

b. 4 3/4 =   4 21/28
    2 9/14 =   2 18/28
$\underline{+\ \ \ 6/7\ \ =\ \ \ \ \underline{24/28}}$
            6 63/28 = 8 7/28 = 8 1/4

c.   4/5 =   32/40
      1 3/4 =   1 30/40
      3 5/8 =   3 25/40
$\underline{+\ \ 7/10\ =\ \ \ \ \underline{28/40}}$
            4 115/40 = 6 35/40 = 6 7/8

9.
Add the following fractions and mixed numbers. Write the answers as fractions or mixed numbers in lowest terms.

a. $6\dfrac{7}{10}$

$+2\dfrac{11}{15}$

b. $4\dfrac{5}{8}$

$5\dfrac{2}{9}$

$+1\dfrac{7}{10}$

c. $5\dfrac{4}{9}$

$6\dfrac{11}{12}$

$1\dfrac{5}{18}$

$+2\dfrac{5}{6}$

SOLUTION:

a. $\quad$ 6 7/10 $\;=\;$ 6 21/30
    $\underline{+\,2\,11/15 \;=\; 2\,22/30}$
    $\qquad\qquad\;\;$ 8 43/30 = 9 13/30

b. $\quad$ 4 5/8 $\;=\;$ 4 225/360
    $\qquad$ 5 2/9 $\;=\;$ 5 80/360
    $\underline{+\,1\,7/10 \;=\; 1\,252/360}$
    $\qquad\qquad\;$ 10 557/360 = 11 197/360

c. $\quad$ 5 4/9 $\;=\;$ 5 16/36
    $\qquad$ 6 11/12 $\;=\;$ 6 33/36
    $\qquad$ 1 5/18 $\;=\;$ 1 10/36
    $\underline{+\,2\;\;5/6 \;=\; 2\,30/36}$
    $\qquad\qquad$ 14 89/36 = 16 17/36

LEARNING OBJECTIVES 2, 4

10.
Subtract the following fractions or mixed numbers. Write the answers as fractions or mixed numbers in lowest terms.

a. $\dfrac{3}{5}$

$-\dfrac{3}{8}$

b. $\dfrac{5}{6}$

$-\dfrac{2}{5}$

SOLUTION:

a. $\quad$ 3/5 $\;=\;$ $\quad$ 24/40
    $\underline{-\,3/8 \;=\; -\,15/40}$
    $\qquad\qquad\qquad$ 9/40

b. $\quad$ 5/6 $\;=\;$ $\quad$ 25/30
    $\underline{-\,2/5 \;=\; -\,12/30}$
    $\qquad\qquad\qquad$ 13/30

11.
Subtract the following fractions or mixed numbers. Write the answers as fractions or mixed numbers in lowest terms.

a. $\dfrac{7}{12}$

$-\dfrac{3}{10}$

b. $\dfrac{3}{4}$

$-\dfrac{1}{6}$

SOLUTION:

a.   7/12 =   35/60
      − 3/10 = − 18/60
                      17/60

b.   3/4 =   9/12
      − 1/6 = − 2/12
                      7/12

12.
Subtract the following fractions or mixed numbers. Write the answers as fractions or mixed numbers in lowest terms.

a. $1\dfrac{5}{18}$

$-\dfrac{17}{27}$

b. $3\dfrac{4}{21}$

$-\dfrac{9}{14}$

SOLUTION:

a.   1 5/18 =   1 15/54 =   69/54
     − 17/27 = − 34/54 = − 34/54
                                         35/54

b.   3 4/21 =   3 8/42 =   2 50/42
     − 9/14 = − 27/42 = − 27/42
                                       2 23/42

13.
Subtract the following fractions or mixed numbers. Write the answers as fractions or mixed numbers in lowest terms.

a. $2\dfrac{7}{16}$

$-\dfrac{9}{10}$

b. $2\dfrac{4}{7}$

$-\dfrac{4}{5}$

SOLUTION:

a.   2 7/16 =   2 35/80 = 1 115/80
     − 9/10 = − 72/80 = − 72/80
                                      1 43/80

b.   2 4/7 =   2 20/35 = 1 55/35
     − 4/5 = − 28/35 = − 28/35
                                    1 27/35

14.
Subtract the following fractions or mixed numbers. Write the answers as fractions or mixed numbers in lowest terms.

a. $3\dfrac{1}{9}$

   $-1\dfrac{5}{12}$

b. $5\dfrac{2}{3}$

   $-3\dfrac{4}{5}$

SOLUTION:

a. $\begin{array}{rcccl} 3\ 1/9 & = & 3\ 4/36 & = & 2\ 40/36 \\ -\ 1\ 5/12 & = & -\ 1\ 15/36 & = & -\ 1\ 15/36 \\ & & & & 1\ 25/36 \end{array}$

b. $\begin{array}{rcccl} 5\ 2/3 & = & 5\ 10/15 & = & 4\ 25/15 \\ -\ 3\ 4/5 & = & -\ 3\ 12/15 & = & -\ 3\ 12/15 \\ & & & & 1\ 13/15 \end{array}$

15.
Subtract the following fractions or mixed numbers. Write the answers as fractions or mixed numbers in lowest terms.

a. $5\dfrac{1}{4}$

   $-\ 4\dfrac{2}{3}$

b. $6\dfrac{3}{8}$

   $-2\dfrac{13}{20}$

SOLUTION:

a. $\begin{array}{rcccl} 5\ 1/4 & = & 5\ 3/12 & = & 4\ 15/12 \\ -\ 4\ 2/3 & = & -\ 4\ 8/12 & = & -\ 4\ 8/12 \\ & & & & 7/12 \end{array}$

b. $\begin{array}{rcccl} 6\ 3/8 & = & 6\ 15/40 & = & 5\ 55/40 \\ -\ 2\ 13/20 & = & -\ 2\ 26/40 & = & -\ 2\ 26/40 \\ & & & & 3\ 29/40 \end{array}$

LEARNING OBJECTIVE 3

16.
A farmer planted 3 2/3 acres of oats, 5 3/4 acres of wheat, and 2 5/6 acres of barley. Compute the total number of acres planted.

SOLUTION:

$\begin{array}{rcl} 3\ 2/3 & = & 3\ 8/12 \\ 5\ 3/4 & = & 5\ 9/12 \\ +\ 2\ 5/6 & = & 2\ 10/12 \\ 10\ 27/12 & = & 12\ 3/12\ \text{acres} \ = \ 12\ 1/4\ \text{acres} \end{array}$

LEARNING OBJECTIVE 4

17.

A rancher planned to fence 5 3/4 acres of land to grow some experimental grasses. After his crew had completed 3 1/3 of the work, it started snowing. How much of the land remained to be fenced?

SOLUTION:

$$
\begin{array}{rcl}
5\ 3/4 & = & 5\ 9/12 \\
-\ 3\ 1/3 & = & -\ 3\ 4/12 \\
\hline
& & 2\ 5/12 \text{ acres}
\end{array}
$$

LEARNING OBJECTIVE 3

18.

Last week, an organic produce farm delivered 11 3/8 pounds of green leaf lettuce, 13 7/16 pounds of romaine lettuce, and 9 5/6 pounds of red leaf lettuce. What is the total weight of the lettuce production?

SOLUTION:

$$
\begin{array}{rcl}
11\ \ 3/8 & = & 11\ \ 18/48 \\
13\ 7/16 & = & 13\ \ 21/48 \\
+\ 9\ \ 5/6 & = & \ \ 9\ \ 40/48 \\
\hline
& & 43\ \ 79/48 = 34\ 31/48 \text{ pounds}
\end{array}
$$

LEARNING OBJECTIVE 4

19.

A husband and wife work at the same company, and decide to walk to work. He leaves home at 7:30 am, and arrives in 12 1/3 minutes. She leaves at 7:45 am, and arrives in 10 3/5 minutes. Compute how much longer it takes him to walk to work than her.

SOLUTION:

$$
\begin{array}{rcl}
12\ 1/3 & = & 12\ 5/15 \ = 11\ 20/15 \\
-\ 10\ 3/5 & = & 10\ 9/15 \ = 10\ \ \ 9/15 \\
\hline
& & 1\ 11/15 \text{ minutes}
\end{array}
$$

LEARNING OBJECTIVE 4

20.

A merchant bought 15 2/5 pounds of apples and pears to sell at a fruit stand. 8 3/4 pounds were apples. Calculate the number of pounds of pears.

SOLUTION:

$$
\begin{array}{rcl}
15\ 2/5 & = & 15\ \ 8/20 \ = 14\ 28/20 \\
-\ 8\ 3/4 & = & \ 8\ 15/20 \ = \ 8\ 15/20 \\
\hline
& & 6\ 13/20 \text{ pounds}
\end{array}
$$

Learning Objective 3

21.
A cyclist rode her bicycle 10 1/2 miles on Monday, 12 7/8 miles on Tuesday, and 11 4/5 miles on Wednesday. Compute the total distance for the three days.

SOLUTION:

```
      10 1/2   = 10 20/40
      12 7/8   = 12 35/40
    + 11 4/5   = 11 32/40
    33 87/40   = 35 7/40 miles
```

Learning Objective 4

22.
A hiking club planned to hike from a valley to a waterfall which were 10 2/5 miles apart. When the hikers stopped for lunch they had already walked 6 3/4 miles. How far were they from the waterfall?

SOLUTION:

```
    10 2/5   =  10  8/20  = 9 28/20
  –  6 3/4   = – 6 15/20  = 6 15/20
                            3 13/20 miles
```

Learning Objectives 3, 4

23.
An electrician bought a 100-foot roll of 12-gauge copper wire, and used 32 7/12 feet for one job and 26 3/4 feet for a second job. How many feet of wire remained on the roll?

SOLUTION:

```
    32 7/12   = 32  7/12
  + 26 3/ 4   = 26  9/12
    58 16/12  = 59 1/3        100 – 59 1/3 = 40 2/3 feet
```

Learning Objective 3

24.
A crafter makes jewelry for friends and relatives. Last week the crafter needed 3/4 ounce of gold and 2 1/6 ounces of silver. Compute the number of ounces needed for the week.

SOLUTION:

```
      3/4   =      9/12
  + 2 1/6   =  2  2/12
               2 11/12 ounces
```

25.

A crafter makes jewelry to sell at crafts fairs. For Christmas, the crafter bought 5 7/10 ounces of gold, 4 4/5 ounces of silver, and 7 5/6 ounces of turquoise. What was the total weight of purchases?

SOLUTION:

$$
\begin{array}{rl}
5\ 7/10 & =\ 5\ 21/30 \\
4\ 4/5 & =\ 4\ 24/30 \\
+\ 7\ 5/6 & =\ 7\ 25/30 \\
\hline
16\ 70/30 & =\ 18\ 10/30 = 18\ 1/3 \text{ ounces (or 1 lb. 2 1/3 oz.)}
\end{array}
$$

LEARNING OBJECTIVE 4

26.

Two chefs agree to provide 3 1/2 gallons of soup for a benefit luncheon.
One chef brings 1 7/8 gallons. Compute how much the other chef must bring.

SOLUTION:

$$
\begin{array}{rlll}
3\ 1/2 & =\ 3\ 4/8 & =\ 2\ 12/8 \\
-\ 1\ 7/8 & =\ 1\ 7/8 & =\ 1\ 7/8 \\
\hline
& & \ \ \ 1\ 5/8 \text{ gallons}
\end{array}
$$

LEARNING OBJECTIVES 3, 4

27.

Four workers contracted to clean an acre of land. One worker did 1/3 of the job: another did 1/4 of the job and the third did 1/6 of the job. How much of the job did the fourth worker do?

SOLUTION:

$$
\begin{array}{rll}
1/3 & =\ 4/12 \\
1/4 & =\ 3/12 \\
+\ 1/6 & =\ 2/12 \\
\hline
& \ \ 9/12 = 3/4
\end{array}
\qquad
\begin{array}{rll}
1 & =\ 4/4 \\
-\ 3/4 & =\ 3/4 \\
\hline
& \ \ 1/4 \text{ of the job}
\end{array}
$$

28.

A business contracts to plant several hundred plants for the landscape of a major apartment complex. During two weeks, one employee planted 2/5 of the plants, and another employee planted 3/8. What fraction of the plants remain to be planted?

SOLUTION:

$$
\begin{array}{rl}
2/5 & =\ 16/40 \\
+\ 3/8 & =\ 15/40 \\
\hline
31/40 & \quad 1 - 31/40 = 9/40 \text{ of the plants remain}
\end{array}
$$

LEARNING OBJECTIVE 4

29.
A package of veal weighs 5 3/8 ounces. A similar package of pork weighs 4 3/5 ounces. What is the difference in the weight of the two packages?

SOLUTION:

```
   5 3/8  =   5 15/40 =   4 55/40
  – 4 3/5  =   4 24/40 =   4 24/40
                            31/40 ounce
```

LEARNING OBJECTIVE 3

30.
A small candy company specializes in handmade chocolates. In one hour, one employee made 7 1/3 pounds of chocolates, another employee made 8 4/5 pounds, and a third employee made 7 1/4 pounds. Compute the total number of pounds the three employees made together.

SOLUTION:

```
   7 1/3 =  7 20/60
   8 4/5 =  8 48/60
 + 7 1/4 =  7 15/60
             22 83/60 = 23 23/60 pounds
```

31.
A jeweler used 2/3 ounce of gold for earrings and 1 3/4 for a bracelet. Compute the total number of ounces of gold that she used.

SOLUTION:

```
    2/3  =    8/12
  + 1 3/4 =  1 9/12
               1 17/12 = 2 5/12 ounces
```

LEARNING OBJECTIVE 4

32.
It is estimated that a small pickup truck could hold 2/3 cord of firewood. In reality, the truck carries 5/9 cord. Calculate the difference between the estimate and the actual load.

SOLUTION:

```
    2/3  = 6/9
  – 5/9  = 5/9
            1/9 cord
```

LEARNING OBJECTIVES 2, 5

33.
Multiply the following fractions. Reduce products to lowest terms.

a. $\dfrac{3}{4} \times \dfrac{4}{5}$

b. $\dfrac{8}{15} \times \dfrac{5}{12}$

c. $\dfrac{7}{15} \times 45$

SOLUTION:
a.   3/5

b.   2/9

c.   21

34.
Multiply the following fractions. Reduce products to lowest terms.

a. $\dfrac{9}{10} \times \dfrac{5}{18}$

b. $\dfrac{10}{21} \times \dfrac{14}{15}$

c. $\dfrac{6}{11} \times \dfrac{22}{27}$

SOLUTION:
a.   1/4

b.   4/9

c.   4/9

35.
Multiply the following fractions. Reduce products to lowest terms.

a. $\dfrac{1}{2} \times \dfrac{3}{4} \times \dfrac{2}{3}$

b. $\dfrac{3}{8} \times \dfrac{2}{3} \times \dfrac{2}{5}$

c. $\dfrac{3}{5} \times \dfrac{1}{2} \times \dfrac{2}{3}$

SOLUTION:
a.   1/4

b.   1/10

c.   1/5

36.
Multiply the following fractions. Reduce products to lowest terms.

a. $\dfrac{5}{6} \times 36 \times \dfrac{4}{15}$

b. $50 \times \dfrac{4}{5} \times \dfrac{3}{16}$

f. $\dfrac{12}{25} \times \dfrac{2}{27} \times 15$

SOLUTION:
a.   8

b.   7 1/2

f.   8/15

37.
Change the mixed numbers to improper fractions and multiply. Write the answers as mixed numbers in lowest terms.

a. $1\dfrac{1}{6} \times 2\dfrac{2}{5}$

b. $2\dfrac{2}{3} \times 4\dfrac{1}{2}$

c. $1\dfrac{1}{14} \times 2\dfrac{1}{10}$

SOLUTION:
a. 7/6 × 12/5 = 14/5 = 2 4/5   b. 8/3 × 9/2 = 12/1 = 12
c. 15/14 × 21/10 = 2 1/4

38.

Change the mixed numbers to improper fractions and multiply. Write the answers as mixed numbers in lowest terms.

a. $2\frac{1}{3} \times 3\frac{3}{4}$ b. $1\frac{1}{15} \times 2\frac{1}{12}$ c. $3\frac{5}{8} \times 3\frac{1}{3}$

SOLUTION:

a. $7/3 \times 15/4 \ = \ 35/4 = 8\ 3/4$ b. $16/15 \times 25/12 = 20/9 = 2\ 2/9$
c. $29/8 \times 10/3 \ = \ 145/1\backslash2 = 12\ 1/12$

39.

Change the mixed numbers to improper fractions and multiply. Write the answers as mixed numbers in lowest terms.

a. $2\frac{2}{3} \times 6\frac{1}{4} \times 3\frac{3}{5}$ b. $2\frac{1}{4} \times 2\frac{4}{5} \times \frac{3}{7}$ c. $1\frac{7}{8} \times 1\frac{3}{14} \times 1\frac{13}{15}$

SOLUTION:
a. $8/3 \times 25/4 \times 18/5 = 60$ b. $9/4 \times 14/5 \times 3/7 = 27/10 = 2\ 7/10$
c. $15/8 \times 17/14 \times 28/15 = 17/4 = 4\ 1/4$

40.

Change the mixed numbers to improper fractions and multiply. Write the answers as mixed numbers in lowest terms.

a. $1\frac{5}{7} \times \frac{14}{15} \times 3\frac{3}{4}$ b. $1\frac{1}{2} \times 1\frac{1}{3} \times 1\frac{1}{4}$ c. $4\frac{1}{6} \times \frac{8}{15} \times 2\frac{1}{2}$

SOLUTION:
a. $12/7 \times 14/15 \times 15/4 = 6$ b. $3/2 \times 4/3 \times 5/4 = 5/2 = 2\ 1/2$
c. $25/6 \times 8/15 \times 5/2 = 5\ 5/9$

LEARNING OBJECTIVE 6

41.

Divide the following fractions. Write the answers as fractions or mixed numbers in lowest terms.

a. $\frac{2}{3} \div \frac{4}{5}$ b. $\frac{4}{5} \div \frac{2}{3}$ c. $\frac{3}{4} \div \frac{5}{8}$

SOLUTION:
a. $2/3 \times 5/4 = 5/6$ b. $4/5 \times 3/2 = 6/5 = 1\ 1/5$
c. $3/4 \times 8/5 = 6/5 = 1\ 1/5$

42.
Divide the following fractions. Write the answers as fractions or mixed numbers in lowest terms.

a. $\dfrac{5}{8} \div \dfrac{3}{4}$ 　　　　 b. $\dfrac{4}{5} \div 2$ 　　　　 c. $\dfrac{3}{5} \div 6$

SOLUTION:
a. 5/8 × 4/3 = 5/6 　　　　 b. 4/5 × 1/2 = 2/5
c. 3/5 × 1/6 = 1/10

43.
Divide the following fractions. Write the answers as fractions or mixed numbers in lowest terms.

a. $\dfrac{9}{16} \div \dfrac{3}{4}$ 　　　　 b. $\dfrac{3}{4} \div \dfrac{9}{16}$ 　　　　 c. $6 \div \dfrac{4}{5}$

SOLUTION:
a. 9/16 × 4/3 = 3/4 　　　　 b. 3/4 × 16/9 = 4/3 = 1 1/3
c. 6 × 5/4 = 15/2 = 7 1/2

44.
Divide the following fractions. Write the answers as fractions or mixed numbers in lowest terms.

d. $\dfrac{5}{24} \div \dfrac{15}{32}$ 　　　　 e. $\dfrac{12}{35} \div \dfrac{9}{25}$ 　　　　 f. $\dfrac{8}{15} \div \dfrac{4}{5}$

SOLUTION:
d. 5/24 × 32/15 = 4/9 　　　　 e. 12/35 × 25/9 = 20/21
f. 8/15 × 5/4 = 2/3

45.
Change the mixed numbers to improper fractions and divide. Write the answers as fractions or mixed numbers in lowest terms.

a. $1\dfrac{2}{3} \div 3\dfrac{3}{4}$ 　　　　 b. $2\dfrac{1}{4} \div \dfrac{3}{4}$

SOLUTION:
a. 5/3 ÷ 15/4 = 5/3 × 4/15 = 4/9 　　　　 b. 9/4 ÷ 3/4 = 9/4 × 4/3 = 3

46.
Change the mixed numbers to improper fractions and divide. Write the answers as fractions or mixed numbers in lowest terms.

a. $\dfrac{2}{3} \div 1\dfrac{1}{9}$ 　　　　 b. $2\dfrac{1}{2} \div 3\dfrac{3}{4}$

SOLUTION:
a. 2/3 ÷ 10/9 = 2/3 × 9/10 = 3/5 　　　　 b. 5/2 ÷ 15/4 = 5/2 × 4/15 = 2/3

47.

Change the mixed numbers to improper fractions and divide. Write the answers as fractions or mixed numbers in lowest terms.

a. $\quad 1\dfrac{5}{16} \div 1\dfrac{7}{8}$

b. $\quad 6\dfrac{4}{5} \div 4$

SOLUTION:
a. $21/16 \div 15/8 = 21/16 \times 8/15 = 7/10$
b. $34/5 \div 4 = 34/5 \times 1/4 = 17/10 = 1\ 7/10$

48.

Change the mixed numbers to improper fractions and divide. Write the answers as fractions or mixed numbers in lowest terms.

a. $\quad 2\dfrac{2}{5} \div 1\dfrac{1}{15}$

b. $\quad 4\dfrac{5}{6} \div 1\dfrac{1}{2}$

SOLUTION:
a. $12/5 \div 16/15 = 12/5 \times 15/16 = 9/4 = 2\ 1/4$
b. $29/6 \div 3/2 = 29/6 \times 2/3 = 29/9 = 3\ 2/9$

LEARNING OBJECTIVE 5

49.

A small herb garden is 3 1/3 feet by 7 1/2 feet. If there are 9 square feet in 1 square yard, compute the number of square yards that are in the herb garden.

SOLUTION:
$3\ 1/3 \times 7\ 1/2 = 10/3 \times 15/2 = 5/1 \times 5/1 = 25$ square feet;
$25 \div 9 = 2\ 7/9$ square yards

50.

A cook has a recipe for a pasta sauce that calls for 3/4 cup of chopped onions. One recipe makes enough for six servings. Compute the volume of chopped onions that would be required for fifteen servings. (Hint: That means the cook must prepare 15/6 = 2 1/2 recipes.)

SOLUTION:
$3/4 \times 2\ 1/2 = 3/4 \times 5/2 = 15/18 = 1\ 7/8$ cups

51.

An amateur cook entered a "chili cook-off." Among other spices the cook included 1 2/3 teaspoons of cayenne pepper. The cook decided the chili was too hot, and wanted to use only 3/4 as much cayenne as had been used. Compute the new amount of cayenne that would be used.

SOLUTION:
$3/4 \times 1\ 2/3 = 3/4 \times 5/3 = 5/4 = 1\ 1/4$ teaspoons

52.

A painter mixing paint for a client starts with 6 1/2 quarts of white base paint. To each quart, the painter must add 2 2/3 ounces of tint. How many ounces of tint are required for this project?

SOLUTION:
6 1/2 × 2 2/3 = 13/2 × 8/3 = 13/1 × 4/3 = 52/3 = 17 1/3 ounces

53.

A person prepares a rusty metal fence for painting, completing 4 1/2 feet of fence per hour. When there are 2 hours and 20 minutes left to work in the day, how many feet can be prepared during the remainder of the day? (Hint: 20 minutes is 1/3 hour.)

SOLUTION:
2 1/3 × 4 1/2 = 7/3 × 9/2 = 7/1 × 3/2 = 21/2 = 10 1/2 feet

LEARNING OBJECTIVE 6

54.

For her semester project, an art student was required to complete an oil painting that was at least 5 square feet in area. If the student wanted to make a painting that was 3 3/4 feet wide, what is the minimum height that it could be?

SOLUTION:
5 ÷ 3 3/4 = 5 ÷ 15/4 = 5/1 × 4/15 = 4/3 = 1 1/3 feet

LEARNING OBJECTIVE 5

55.

A homeowner contracts with a landscaper to resod the front lawn which is 10 1/2 yards long by 6 2/3 yards wide. Compute the cost if the landscaper charges $8.00 per square yard for labor and materials.

SOLUTION:
10 1/2 × 6 2/3 = 21/2 × 20/3 = 70 sq. yd; 70 × $8 = $560

56.

A car manufacturer claims that one of its new models will travel 32 miles on one gallon of gasoline. If the claim is true, how far should it travel on 3 3/4 gallons of gasoline?

SOLUTION:
32 × 3 3/4 = 32/1 × 15/4 = 8/1 × 15/1 = 120 miles

LEARNING OBJECTIVES 3, 4, 5

57.
A house painter uses about 2 1/2 quarts of paint per hour on a specific type of wooden exterior trim. For an 8-hour day, the painter's preparation work is 3/4 of an hour, with 1/2 hour for cleanup. If the person paints the remainder of the 8 hours, compute the number of quarts of paint that will be used.

SOLUTION:
3/4 + 1/2 = 3/4 + 2/4 = 5/4 = 1 1/4 hours;
8 – 1 1/4 = 7 4/4 – 1 1/4 = 6 3/4 hours;
6 3/4 × 2 1/2 = 27/4 × 5/2 = 135/8 = 16 7/8 quarts

LEARNING OBJECTIVES 5, 6

58.
A school cafeteria prepares soup in five-gallon stock pots. Because of waste, the cooks only get 3 3/4 gallons of soups from one pot. If one large serving measures about 1 1/3 cups, how many large servings can they get out of one pot? (One gallon = 4 quarts = 16 cups.)

SOLUTION:
1 gal = 16 cups 3 3/4 × 16 = 60 cups;
60 ÷ 1 1/3 = 60 ÷ 4/3 = 60/1 × 3/4 = 45 servings

LEARNING OBJECTIVE 5

59.
In the bedroom of a house, there was 15/16 inch of space underneath the door before a carpet and pad were installed. After the installation, the space was only 2/5 of the previous space. Find the space under the door after the installation.

SOLUTION:
2/5 × 15/16 = 3/8 inch

60.
A lawn mower manufacturer claims that one of its new gas-powered lawn mowers will run for 1 1/3 hour on 1 quart of gasoline. If the claim is true, how long should this lawn mower run on 2 1/2 quarts of gasoline?

SOLUTION:
1 1/3 × 2 1/2 = 4/3 × 5/2 = 10/3 = 3 1/3 hours

LEARNING OBJECTIVE 6

61.
A renter wanted to paint a small wall in her apartment. She bought a can paint that was supposed to cover 110 square feet. The height of the wall was 8 1/3 feet high. What length of wall could she paint with this one can of paint?

SOLUTION:
110 ÷ 8 1/3 = 110 ÷ 25/3 = 110/1 × 3/25 = 66/5 = 13 1/5 feet

LEARNING OBJECTIVE 5

62.

A concrete walkway will be 18 feet long, 3 2/3 feet wide and 1/3 foot thick. Compute the number of cubic feet of concrete used to make the patio.

SOLUTION:
18 × 3 2/3 × 1/3 = 18/1 × 11/3 × 1/3 = 2/1 × 11/1 × 1/1 = 22 cubic feet

LEARNING OBJECTIVES 5, 6

63.

An accountant often eats lunch at a restaurant that is 3/8 mile from her office. She makes the walk in 7 1/2 minutes. At that rate, how long will it take her to walk the two miles from her home to her office?

SOLUTION:
2 ÷ 3/8 = 2/1 × 8/3 = 16/3; 16/3 × 7 1/2 = 16/3 × 15/2 = 40 minutes; or
7 1/2 ÷ 3/8 = 15/2 × 8/3 = 20 minutes per mile;
20 × 2 = 40 minutes

LEARNING OBJECTIVES 3, 5

64.

Many corporate stocks are priced below $1 dollar per share. Such stocks are sometimes called "penny stocks." If an investor bought the stock of a biotechnical company for $ 5/8 per share, compute the new price per share if the amount of the price increase is 1 1/5 of the original price.

SOLUTION:
increase = 1 1/5 × $ 5/8 = 6/5 × 5/8 = $ 3/4
new price = $ 5/8 + $ 3/4 = $ 5/8 + $ 6/8 = $ 11/8 = $1 3/8

LEARNING OBJECTIVE 6

65.

The employee who applies the finishing touches and final adjustments to bicycles takes about 1 1/3 hours per bicycles. Calculate the total number of bicycles that he can completely finish in an 8-hour day.

SOLUTION:
8 ÷ 1 1/3 = 8/1 ÷ 4/3 = 8/1 × 3/4 = 6 bicycles

Chapter 3 DECIMALS

PROBLEMS

LEARNING OBJECTIVE 1

1.
Use digits to write each number that is expressed in words.

a.   Seventeen and sixteen thousandths _____
b.   Seven and twenty-five thousandths _____
c.   Four hundred eighty-eight ten-thousandths _____

SOLUTION:
a.   17.016   b.   7.025   c.   0.0488

2.
Use digits to write each number that is expressed in words.

a.   Thirty-five thousandths _____
b.   Four hundred four and six hundredths _____
c.   Five thousand two hundred-thousandths _____

SOLUTION:
a.   0.035   b.   404.06   c.   0.05002

3.
Use words to write each number that is expressed in digits.

a.   4.284 _____
b.   32.51 _____
c.   6.099 _____

SOLUTION:
a.   four and two hundred eighty-four thousandths
b.   thirty-two and fifty-one hundredths
c.   six and ninety-nine thousandths

4.
Use words to write each number that is expressed in digits.

a.   0.0012 _____
b.   12.7344 _____
c.   4.00961 _____

SOLUTION:
a.   twelve ten-thousandths
b.   twelve and seven thousand three hundred forty-four ten-thousandths
c.   four and nine hundred sixty-one hundred-thousandths

LEARNING OBJECTIVE 2

5.
Round each monetary amount to the nearest cent; round the non-monetary numbers to the nearest thousandth.

a. $41.875 _____
b. $1.2749 _____
c.   0.07351 ounces _____
d.   0.22499 feet _____
e.   4.099489 pounds _____
f. $0.44501 _____

SOLUTION:
a. $41.88   b. $1.27   c. 0.074 ounces
d.   0.225 feet   e. 4.099 pounds   f. $0.45

6.
Round each monetary amount to the nearest cent; round the nonmonetary numbers to the nearest thousandth.

a. $0.24499 _____
b. $2.385 _____
c.   0.69164 pounds _____
d.   2.63151 gallons _____
e.   2.375388 feet _____

SOLUTION:
a. $0.24   b. $2.39   c. 0.692 pounds
d.   2.632 gallons   e. 2.375 feet

LEARNING OBJECTIVE 3

7.
Add the following decimal numbers.
a.   0.885   b.   0.146   c.   7.006
      0.39   1.7092   3.47
    + 0.0053   + 0.0045   + 18.8835

SOLUTION:
a.   1.2803   b.   1.8597   c.   29.3595

8.

Add the following decimal numbers.

a.	36.7484	b.	904.98	c.	0.055
	590.28		72.5772		4.56
	+ 4.1763		+ 2,404.115		+39.7468

SOLUTION:

a.   631.2047   b.   3,381.6722   c.   44.3618

9.

Add the following decimal numbers.

a.	0.854	b.	0.85	c. 12.464
	0.86		0.3534	0.295
	+ 0.3528		+ 0.688	+ 13.6579

SOLUTION:

a.   2.0668   b.   1.8914   c. 26.4169

10.

Add the following decimal numbers.

a.	49.8715	b.	444.92	c.	0.07
	801.97		75.0886		1.283
	+ 48.4338		+ 2,500.		+ 0.93

SOLUTION:

a.   900.2753   b.   3,020.0086   c.   2.283

LEARNING OBJECTIVE 4

11.

Subtract the following decimal numbers.

a.	4.5051	b.	0.724	c.	27.58103
	− 0.31747		− 0.4681		− 19.797

SOLUTION:

a.   4.18763   b.   0.2559   c.   7.78403

12.

Subtract the following decimal numbers.

a.	414.02	b.	6,000.	c.	2.101
	− 175.624		− 197.462		− 1.898

SOLUTION:

a.   238.396   b.5,802.538   c.   0.203

13.
Subtract the following decimal numbers.

a.   1.00425   b.   0.37   c.   84.34475
      − 0.32559   − 0.2206   − 39.667

SOLUTION:
a.   0.67866   b.   0.1494   c.   44.67775

14.
Subtract the following decimal numbers.

d.   207.011   e.   5,000.   f.   22.021
     − 139.0125   − 1,500.25   − 6.45123

SOLUTION:
d.   67.9985   e.   3,499.75   f.   15.56977

LEARNING OBJECTIVES 3, 4

15.
Three boxes of pears weighing 32.4, 33.8 and 33.4 pounds were shipped. Compute the total weight.

SOLUTION:
32.4 + 33.8 + 33.4 = 99.6 pounds

LEARNING OBJECTIVE 3

16.
An electrical contractor started the day with 284.2 feet of 10 gauge copper wire. He used 42.5 feet on one job and 114.8 feet on another job. How many feet of wire did he have at the end of the day?

SOLUTION:
42.5 + 114.8 = 157.3; 284.2 − 157.3 = 126.9 feet

LEARNING OBJECTIVES 3, 4

17.
A restaurant had 18.6 pounds of pork on Wednesday morning and received 20.9 pounds Wednesday. On Thursday morning it had 9.8 pounds on hand. How many pounds did it use on Wednesday?

SOLUTION:
18.6 + 20.9 = 39.5 pounds; 39.5 − 9.8 = 29.7 pounds

LEARNING OBJECTIVE 3

18.
A salesperson drives 74.9 miles on Monday, 59.8 on Tuesday, 65.5 on Wednesday, and 86.4 on Thursday. On Friday the salesperson stayed home. What was the total distance traveled last week?

SOLUTION:
74.9 + 59.8 + 65.5 + 86.4 = 286.6 miles

LEARNING OBJECTIVES 3, 4

19.
A grocery store had 76.4 pounds of chicken in refrigeration on Friday morning. During the day, customers purchased 59.8 pounds, and 8.8 pounds were waste and thrown away. Calculate the number of pounds that were left on Friday night.

SOLUTION:
59.8 + 8.8 = 68.6 pounds gone; 76.4 – 68.6 = 7.8 pounds remaining

LEARNING OBJECTIVE 3

20.
A production engineer wanted to know how long it should take to make metal rods with a lathe. Four rods were made, and the time was recorded. The results were 28.5 seconds, 29.2 seconds, 31.8 seconds, and 29.7 seconds. Compute the total time to make all four rods.

SOLUTION:
28.5 + 29.2 + 31.8 + 29.7 = 119.2 seconds

21.
A fresh produce wholesaler shipped 145.6 pounds of apples, 88.3 pounds of pears, and 63.7 pounds of plums to three small neighborhood grocery stores. What was the total weight of the fruit shipped?

SOLUTION:
145.6 + 88.3 + 63.7 = 297.6 pounds

22.
On April 12, a tax accountant took six tax returns to the post office. They weighed 4.2, 7.7, 8.5, 6.3 8.1, and 6.3 ounces. Determine the total weight in ounces.

SOLUTION:
4.2 + 7.7 + 8.5 + 6.3 + 8.1 + 6.3 = 41.1 ounces

LEARNING OBJECTIVES 3, 4

23.

A pet store had $468.42 cash on hand. It received cash payments of $62.88 and $59.14. It paid out $56.50 to have the windows washed. Determine the amount of cash the pet store had left.

SOLUTION:
$468.42 + $62.88 + $59.14 − $56.50 = $533.94

24.

A pharmacy started the month with $124.57 worth of dental floss. During the month, it received dental floss worth $42.44 and sold dental floss worth $89.95. Compute the value of the remaining dental floss.

SOLUTION:
$124.57 + $42.44 = $167.01; $167.01 − $89.95 = $77.06

25.

A restaurant had $356.87 cash on hand in the morning. Total cash receipts were $873.45 from lunch and $1,462.58 from dinner. The restaurant gave $2,200 cash to a security service at closing time. What was the amount of cash on hand?

SOLUTION:
$356.87 + $873.45 + $1,462.58 = $2,692.90; $2,692.90 − $2,200.00 = $492.90

LEARNING OBJECTIVE 3

26.

A hardware store sells most kinds of nails by the pound. A contractor bought 6.8 pounds of roofing nails, 7.7 pounds of "10-penny" nails, and 8.2 pounds of "8-penny" nails. Compute the total pounds of nails that the contractor bought.

SOLUTION:
6.8 + 7.7 + 8.2 = 22.7 pounds

LEARNING OBJECTIVES 3, 4

27.

A landscaping firm brought three trucks loaded with topsoil to a job site. Two trucks carried 7.75 cubic yards each, and one truck carried 5.25 cubic yards. When the job was finished, 3.5 cubic yards remained. Find the number of cubic yards used.

SOLUTION:
7.75 + 7.75 + 5.25 = 20.75 cu. yds.; 20.75 - 3.50 = 17.25 cubic yards

LEARNING OBJECTIVE 3

28.
A secretary finished a two-page letter in 14.2 minutes. The secretary typed page one in 6.8 minutes and page two in 4.6 minutes. Compute the time that was spent printing, etc. (not typing time).

SOLUTION:
14.2 – 6.8 = 7.4 minutes; 7.4 – 4.6 = 2.8 minutes

LEARNING OBJECTIVES 3, 4

29.
To promote good employee health, the cafeteria at a corporation serves many fresh vegetables. It bought 21.4 pounds of celery, 33.2 pounds of carrots, 8.6 pounds of radishes, 12.8 pounds of broccoli, and 52.6 pounds of lettuce. What was the total weight of the vegetables purchased?

SOLUTION:
21.4 + 33.2 + 8.6 + 12.8 + 52.6 = 128.6 pounds

30.
When it opened on Monday morning, a local delicatessen had 26.8 pounds of salami. During the week, it received a shipment of 84.9 pounds of salami. Also during the week, it used 42.8 pounds of salami in sandwiches and sold 34.2 pounds in bulk to retail customers. How much salami remained at the end of the week?

SOLUTION:
26.8 + 84.9 – 42.8 – 34.2 = 34.7 pounds

31.
On Tuesday, a produce market sold 11.8 pounds of tangerines, 18.3 pounds of oranges and 12.5 pounds of grapefruit. On Saturday, it sold 19.4 pounds of tangerines, 31.7 pounds of oranges and 22.6 pounds of grapefruit. How many more pounds of these fruits did the market sell on Saturday than on Tuesday?

SOLUTION:
11.8 + 18.3 + 12.5 = 42.6 pounds sold on Tuesday
19.4 + 31.7 + 22.6 = 73.7 pounds sold on Saturday
73.7 – 42.6 = 31.1 more pounds sold on Saturday

LEARNING OBJECTIVE 5

32.
Multiply; round off monetary products to the nearest cent. Do not round off the non-monetary products.

a. 5.193×6.2 b. $\$5.20 \times 50.5$ c. 9.486×0.037

SOLUTION:
a.   32.1966   b. \$262.60   c. 0.350982

33.
Multiply; round off monetary products to the nearest cent. Do not round off the non-monetary products.

a. 217.4×0.494

b. $\$76.44 \times 6.7$

c. $\$25.65 \times 4.27$

SOLUTION:
a.   107.3956

b. $512.15

c. $109.53

34.
Multiply; round off monetary products to the nearest cent. Do not round off the non-monetary products.

a. $\$46.82 \times 14.1$

b. 0.625×0.25

c. $\$509.88 \times 7.9$

SOLUTION:
a. $660.16

b.   0.15625

c. $4,028.05

35.
Multiply; round off monetary products to the nearest cent. Do not round off the non-monetary products.

a. 31.402×6.55

b. $\$8.125 \times 400$

c. 16.54×3.93

SOLUTION:
a.   205.6831

b. $3,250

c.   65.0022

36.
Multiply; round off monetary products to the nearest cent. Do not round off the non-monetary products.

a. 6.047×0.0075

b. $\$45.83 \times 21.6$

c. 470.028×0.0906

SOLUTION:
a.   0.0453525

b. $989.93

c.   42.5845368

37.
Multiply; round off monetary products to the nearest cent. Do not round off the non-monetary products.

a. $\$0.625 \times 8,000$

b. 2.9019×0.047

c. $\$27.35 \times 16.75$

SOLUTION:
a. $5,000

b.   0.1363893

c. $458.11

LEARNING OBJECTIVE 6

38.
Divide; round off monetary quotients to the nearest cent; round non-monetary quotients to four decimal places.

a. $\$17.55 \div 7$

b. $13.115 \div 3.28$

c. $\$1.32 \div 0.16$

SOLUTION:
a. $2.51

b.   3.9985

c. $8.25

39.
Divide; round off monetary quotients to the nearest cent; round non-monetary quotients to four decimal places.

a. $4.4868 \div 2.53$

b. $7.52 \div 0.45$

c. $\$154.75 \div 75$

SOLUTION:
a. 1.7734

b. 16.7111

c. $2.06

40.
Divide; round off monetary quotients to the nearest cent; round non-monetary quotients to four decimal places.

a. $0.038 \div 0.007$

b. $\$358.88 \div 11.6$

c. $0.45409 \div 0.649$

SOLUTION:
a. 5.4286

b. $30.94

c. 0.6997

41.
Divide; round off monetary quotients to the nearest cent; round non-monetary quotients to four decimal places.

a. $\$5.92 \div 0.25$

b. $\$1,524.50 \div 310$

c. $6.275 \div 13$

SOLUTION:
a. $23.68

b. $4.92

c. 0.4827

42.
Divide; round off monetary quotients to the nearest cent; round non-monetary quotients to four decimal places.

a. $\$72.63 \div 5.4$

b. $112.25 \div 8.27$

c. $\$306.03 \div 5.05$

SOLUTION:
a. $13.45

b. 13.5732

c. $60.60

43.
Divide; round off monetary quotients to the nearest cent; round non-monetary quotients to four decimal places.

a. $12.6 \div 0.692$

b. $627.17 \div 1.7$

c. $\$12.25 \div 40$

SOLUTION:
a. 18.2081

b. 368.9235

c. $0.31

LEARNING OBJECTIVE 7

44.

Solve the following multiplication and division problems by moving the decimal point to the right or left.

a. $41.00 ÷ 100 = _____

b.   6.34 pints × 1,000 = _____

c.   5,280 feet ÷ 1,000 = _____

d. $15.42 × 10,000 = _____

e.   7.47 yards × 100 = _____

SOLUTION:

a. $0.41   b. 6,340 pints   c. 5.28 feet

d. $154,200   e. 747 yards

45.

Solve the following multiplication and division problems by moving the decimal point to the right or left.

a.   745.6 ounces ÷ 1000 = _____

b. $47.50 × 10 = _____

c.   0.036 gallons × 10,000 = _____

d. $71.50 ÷ 10 = _____

e.   212.75 yards × 100 = _____

SOLUTION:

a.   0.7456 ounces   b. $475   c. 360 gallons

d. $7.15   e. 21,275 yards

LEARNING OBJECTIVE 8

46.

For each of the following multiplication and division problems, determine which estimate is most nearly correct.

a.   0.391 × 81.425   b. 0.0874 × 0.0539   c. 0.30667 × 4.8508
      A) 0.32   A) 0.0045   A) 0.15
      B) 3.2   B) 0.045   B) 1.5
      C) 32   C) 0.45   C) 15
      D) 320   D) 4.5   D) 150

d.   701.47 ÷ 19.15   e. 0.652 ÷ 0.816   f. 0.0000733 ÷ 0.0789
      A) 0.35   A) 0.08   A) 0.00009
      B) 3.5   B) 0.8   B) 0.0009
      C) 35   C) 8   C) 0.009
      D) 350   D) 80   D) 0.09

SOLUTION:

a.   C) 32   b. A) 0.0045   c. B) 1.5

d.   C) 35   e. B) 0.8   f. B) 0.0009

LEARNING OBJECTIVE 5

47.
David's Delicatessen sells macaroni salad for $1.35 for 1/2 pint. Compute the cost of 3.75 quarts (one quart = 2 pints).

SOLUTION:
3.75 qts. × 2 pints per qt. = 7.5 pints; $1.35 × 2 = $2.70 per pint;
7.5 pints × $2.70 = $20.25

48.
Waterfront Restaurant sells "chili-to-go" for $6.95 a quart. Find the cost of 1.25 gallons (one gallon = 4 quarts).

SOLUTION:
1.25 gal × 4 qts per gal = 5 qts; 5 qts × $6.95 per qt = $34.75

49.
Kathy Reynolds, a college student, works as a part-time retail clerk in a clothing store. Kathy can buy clothes at a discount and earns $12.45 per hour. Compute her earnings for a week when she worked 17.25 hours.

SOLUTION:
$12.45 per hour × 17.25 hours = $214.76

LEARNING OBJECTIVES 3, 5

50.
High school student Kevin Parris worked after school for 2.3 hours on Thursday and 4.25 hours on Friday. Calculate the amount that he earned at $7.20 per hour.

SOLUTION:
2.3 + 4.25 = 6.55 hours; 6.55 hours × $7.20 per hour = $47.16

LEARNING OBJECTIVE 6

51.
Eleanor Gunther was paid $76.19 for working 5.75 hours. What is the Eleanor's rate of pay?

SOLUTION:
$76.19 ÷ 5.75 hours = $13.25 per hour

LEARNING OBJECTIVES 3, 5

52.
Oswald Garden Service charges $12.35 per hour per man for general yard maintenance, but charges $17.75 per hour for cement work and tree removal. Compute their total charges for a job which took 5.6 man-hours of general yard maintenance work and 2.4 man-hours of tree removal.

SOLUTION:
5.6 hours × $12.35 per hour = $69.16; 2.4 hours × $17.75 per hour = $42.60;
$69.16 + $42.60 = $111.76

LEARNING OBJECTIVES 5

53.
Betsy's new car travels 36.4 miles on one gallon of gasoline. How far can her car go on 8.25 gallons of gasoline (round to the nearest tenth)?

SOLUTION:
36.4 miles per gallon × 8.25 gallons = 300.3 miles

LEARNING OBJECTIVE 6

54.
Oscar's new pickup truck travels 30.8 miles on one gallon of gasoline. Compute the gallons of gasoline that his truck would use on a 450-mile journey (round to the nearest tenth).

SOLUTION:
450 miles ÷ 30.8 miles per gallon = 14.6 gallons

LEARNING OBJECTIVES 3, 6

55.
The former owner of a used car told the new buyer that the car could travel for 36.4 miles on one gallon of gasoline. The buyer tested the car by driving it for 170 miles on 4.5 gallons of gasoline. Was this better or worse than the claim, and by how many miles per gallon? (Round to the nearest tenth.)

SOLUTION:
170 miles ÷ 4.5 gallons = 37.8 miles per gallon;
37.8 - 36.4 = 1.4 miles per gallon better

LEARNING OBJECTIVE 5

56.
In the winter, red bell peppers are selling for $2.79 per pound. What is the total price of four red peppers which have a combined weight of 2.43 pounds?

SOLUTION:
$2.79 × 2.43 pounds = $6.78

LEARNING OBJECTIVE 6

57.
A 200-foot roll of tubing used primarily in automobiles was recently bought for $35.70. Compute the cost in cents per foot.

SOLUTION:
$35.70 ÷ 200 feet = 17.85 cents per foot

58.
Seventy-five feet of tubing can eventually sell for a total of $48. How much must be charged per foot to earn the $48?

SOLUTION:
$48 ÷ 75 feet = 64 cents per foot

59.
Bill Pierson buys 75 feet of tubing for $18. Bill cuts the tubing in shorter pieces and resells all of it. Bill's total income from the tubing is $63. Compute Bill's profit per foot.

SOLUTION:
$18 ÷ 75 feet = 24 cents per foot; $63 ÷ 75 feet = 84 cents per foot;
Profit = 60 cents per foot

LEARNING OBJECTIVES 3, 5

60.
A certain cut of beef costs $7.59 per pound, and a similar cut of pork costs $5.19 per pound. What is the total cost of 3.25 pounds of the beef and 3.75 pounds of the pork?

SOLUTION:
3.25 pounds × $7.59 per pound = $24.67 for the beef
3.75 pounds × $5.19 per pound = $19.46 for the pork
$24.67 + $19.46 = $44.13 total

LEARNING OBJECTIVE 6

61.
A warehouse store sells a package of 125 steel washers for $2.75. What is the price per washer when they are purchased in this package? (Find the price to the nearest tenth of a cent.)

SOLUTION:
$2.75 ÷ 125 = $0.022 or 2.2 cents per washer.

62.

The wholesale price of a plastic irrigation bubbler is 25 cents. How many plastic bubblers can be purchased for $165?

SOLUTION:
$165 ÷ 25 cents = $165 ÷ $1/4 = 165 × 4/1 = 660 bubblers

63.

Rubber washers are sold for 37.5 cents per dozen, wholesale. Compute the amount that will be charged for 480 dozen washers.

SOLUTION:
480 × 37.5 cents = 480 × $0.375 = $180

64.

Large copper tubing costs 87.5 cents per foot. How much will 1,600 feet of tubing cost?

SOLUTION:
1,600 × 87.5 cents = 1,600 × $0.875 = $1,400

LEARNING OBJECTIVE 6

65.

Julian's City Hardware store sells single strand 12-gauge copper electrical wire at 18 cents per foot. The same wire also comes in a 250-foot roll for $37.49 a roll. At the 18 cents per foot price, how many feet would the customer be able to purchase for $37.49?

SOLUTION:
$37.49 ÷ $0.18 per foot = 208.3 feet

LEARNING OBJECTIVES 3, 5

66.

Seaside Fish Market sells salmon for $7.89 per pound and red snapper for $5.79 per pound. What is the total cost of 2.42 pounds of salmon and 3.16 pounds of red snapper?

SOLUTION:
2.42 pounds × $7.89 per pound = $19.09 for the salmon
3.16 pounds × $5.79 per pound = $18.30 for the red snapper
$19.09 + $18.30 = $37.39 total

67.

Dave Miles earns $8.40 per hour working in a restaurant on weekdays. If Dave works at least 30 hours during the week on weekdays, then he earns $12.60 per hour on the following Saturday. How much would Dave earn during a week in which he worked 34.5 hours during the week and 6.25 more hours on Saturday?

SOLUTION:

34.5 hours × $8.40 per hour = $289.80 during the week

6.25 hours × $12.60 per hour = $78.75 on Saturday

$289.80 + $78.75 = $368.55 total

Chapter 4   WORD PROBLEMS AND EQUATIONS

PROBLEMS

LEARNING OBJECTIVE 1

1.
a.   Add: $7 + 3 + 6 + 4 + 8 + 2 =$ _____
b.   Add: $9 + 1 + 3 + 7 + 8 + 2 + 6 + 4 =$ _____
c.   Subtract: $50 - 8 - 2 - 6 - 4 - 9 - 1 =$ _____
d.   Subtract: $35 - 5 - 3 - 7 - 4 -- 6 =$ _____

SOLUTION:
a.   30
b.   40
c.   20
d.   10

2.
a.   Multiply: $3 \times 2 \times 5 \times 3 =$ _____
b.   Multiply: $4 \times 1 \times 2 \times 8 =$ _____
c.   Divide: $32 \div 2 \div 2 \div 4 =$ _____
d.   Divide: $48 \div 2 \div 3 \div 2 =$ _____

SOLUTION:
a.   90
b.   64
c.   2
d.   4

3.
a.   Mixed: $3 + 3 - 2 + 5 \times 3 =$ _____
b.   Mixed: $7 \times 7 - 9 \times 2 \div 8 =$ _____
c.   Mixed: $7 + 3 - 5 \times 4 + 1 \div 3 + 7 =$ _____
d.   Mixed: $12 \times 2 \, c \, 3 \times 8 + 3 =$ _____

SOLUTION:
a.   27
b.   10
c.   14
d.   67

4.
a.   Mixed: $100 \div 10 \times 30 - 50 \div 5 = $ _____
b.   Mixed: $4 \times 4 + 4 \div 4 - 4 \times 100 = $ _____
c.   Mixed: $3,000 + 3,000 - 1,000 \times 3 = $ _____
d.   Mixed: $8 \div 2 \times 30 + 30 - 25 \div 5 = $ _____

SOLUTIION:
a.   50
b.   100
c.   15,000
d.   25

LEARNING OBJECTIVE 2

5.
A company earned $14,500 in January, $7,600 in February, and $11,700 in March. Compute what the company earned in the three months.

SOLUTION: $33,800

6.
An airplane flew 2,720 miles the first day, 2,740 miles the second day, and 2,980 miles the third day. Compute the number of miles the airplane flew in the three days.

SOLUTION: 8,440

7.
A customer had a twenty-dollar bill and spent $11.93 at the store. Compute the amount the customer had left.

SOLUTION: $8.07

8.
A store owner had 36 carpets and sold 17 of them. Compute the number of carpets the owner had left.

SOLUTION: 19 carpets

9.
Robert Nelson sold 20 Ipods at $20 each and 3 CD's at $15 each. How much was the total sale?

SOLUTION: $445

10.
Robert Nelson's second customer bought 11 CD's at $15 and 4 CD's at $11. How much was the total of the second sale?

SOLUTION: $209

11.

A bus travels 8 miles on a gallon of gas. The bus used 61 gallons of gas last week. Compute the number of miles the bus traveled.

SOLUTION: 488 miles

12.

An employee worked 40 hours at $11.25 per hour. Compute the amount of money the employee earned.

SOLUTION: $450.00

13.

A store owner wanted to give each of her 20 employees an equal share of a special bonus fund containing $6,716. Compute the amount of bonus money each employee received.

SOLUTION: $335.80 bonus

14.

An airplane flew for 80 hours in a week. The airplane used 55 gallons of gas per hour. Compute the number of gallons of gas the airplane used that week.

SOLUTION: 4,400 gallons

15.

A company purchased a desk for $399, a chair for $139, a lamp for $89, and a table for $79. The store offered a discount of $100 for purchases totaling $400 or more. Compute the amount the company paid for the furniture.

SOLUTION: $606

16.

A company earned $19,862 in April and $11,072 in May. In June the company lost $3,990. Compute the amount the company earned during the three-month period.

SOLUTION: $26,944

17.

A company ordered 47 square yards of red carpet, 48 square yards of blue carpet, and 46 square yards of yellow carpet. The carpet cost $20 per square yard including tax. Compute the amount the company paid for the carpeting.

SOLUTION: $2,820

18.

A school purchased 12 gallons of milk on Monday, 14 gallons on Tuesday, 11 gallons on Wednesday, 12 gallons on Thursday, and 14 gallons on Friday. If milk cost $3.25 per gallon, compute the amount the school spent on milk that week.

SOLUTION: $204.75

19.
A school class is washing cars on Saturday to earn $200 for a charity drive. The students collect a donation of $6.50 for each car they wash. The students have washed 28 cars. How many more dollars do they need to earn to reach their goal?

SOLUTION: $18.00 more

20.
An employee earned $12.50 per hour and worked 40 hours. The employer subtracted $56 for income taxes and other necessary deductions. Compute the amount the employee had left to spend.

SOLUTION: $444 left

21.
Twenty-five people paid $20 per ticket for a special movie. The twenty-five people spent an additional $170 for food and drinks at the theater. What was the total expenditure?

SOLUTION: $670

22.
Office Depot sold 28 printer units for $360 each. In addition they sold 71 cartons of paper products at $34.50 each. What was the total sale?

SOLUTION: $12,529.50

23.
A company purchased 26 desks for $240 each and 18 chairs for $59 each. Compute the total purchase price.

SOLUTION: $7,302

24.
A trucking company had 3 trucks. The first truck traveled 790 miles, the second truck traveled 830 miles, and the third truck traveled 948 miles. Each truck traveled 12 miles on a gallon of gas. Gas cost $1.60 per gallon. Compute the amount the trucking company spent on gas.

SOLUTION: $342.40 cost of gas

25.
A store owner planned to give away $1,200 at Christmas. The owner gave $100 to each of the 4 regular employees and $25 to each of the 3 temporary employees. The remaining money was given to a local charity. Compute the amount the local charity received.

SOLUTION: $725 to charity

26.
A customer purchased 3 tablets for $1.25 each, 6 pencils for $0.20 each, and 2 notebooks for $1.35 each. All prices include tax. The customer handed the salesperson a ten-dollar bill. Compute the amount of change the customer should have received.

SOLUTION: $2.35

27.

Last week, a toy manufacturer made the following toys and sold them to a department store at the following prices: 144 dolls at $8 each, 72 drums at $3 each, and 288 train sets at $35 each. The toy manufacturer paid a total of $6,228 for rent, materials, and labor. Compute the amount the toy manufacturer had left after paying all expenses.

SOLUTION: $5,220 money left

28.

A company had November and December sales last year of $45,000 and $55,000, respectively, and November sales this year of $30,000. Compute the amount needed to be sold in December to equal last year's November and December sales.

SOLUTION: $70,000 December sales

29.

The total cost of 5 identical items is $26.75. Compute the cost of one of these items.

SOLUTION: $5.35 per item

30.

At 19 miles per gallon, compute the number of miles a car would run on 16 gallons of gas.

SOLUTION: 304 miles

LEARNING OBJECTIVE 3

31.

Compute the number of miles a person will travel at 65 miles per hour for six hours.

SOLUTION: 390 miles

32.

If a person travels 280 miles in 5 hours, compute the person's miles per hour.

SOLUTION: 56 miles per hour

33.

If a person travels 240 miles in 300 minutes, compute the person's miles per hour.

SOLUTION: 48 miles per hour

34.

At 45 miles per hour, compute the time it would take to travel a total of 540 miles.

SOLUTION: 12 hours

35.

Bob and Mary start traveling toward each other from 600 miles apart. Bob is traveling at 45 miles per hour; Mary at 55 miles per hour. Compute the time elapsed before they meet.

SOLUTION: 6 hours

36.

Bob and Mary start traveling toward each other from 1,050 miles apart. Bob is traveling at 24 miles per hour; Mary at 26 miles per hour. Compute the time elapsed before they meet.

SOLUTION: 21 hours

37.

Bob and Mary start traveling toward each other from 825 miles apart. Bob is traveling at 35 miles per hour and Mary at 40 miles per hour. Compute the distance Bob will travel before they meet.

SOLUTION: 385 miles

38.

Bob and Mary start traveling toward each other from 960 miles apart. Bob is traveling at 40 miles per hour and Mary at 40 miles per hour. Compute the distance Bob will travel before they meet.

SOLUTION: 480 miles

39.

A train leaves San Francisco for Los Angeles at 8:00 am. A second train leaves Los Angeles for San Francisco at 8:00 am the same day. The distance between the two cities is 400 miles. If each train travels at an average speed of 80 miles per hour, what time will it be when the two trains meet.

SOLUTION: 10:30 am

LEARNING OBJECTIVE 4

40.
a.   Mixed: $6 \times 4 \div 2 \times 3 \div 6 =$ _____
b.   Mixed: $18 \div 9 \times 12 \div 3 \div 4 =$ _____
c.   Addition equation: $9 + 4 = 3 +$ _____
d.   Subtraction equation: $41 - 6 = 50 -$ _____
e.   Mixed equation: $12 + 6 = 22 -$ _____

SOLUTION:
a.   6
b.   2
c.   10
d.   15
e.   4

41.

a.   Mixed equation: $39 - 19 = 15 +$ _____

b.   Multiplication equation: $6 \times 4 = 8 \times$ _____

c.   Division equation: $48 \div 3 = 32 \div$ _____

d.   Mixed equation: $56 \div 4 = 2 \times$ _____

e.   Mixed equation: $12 \times 2 = 72 \div$ _____

SOLUTION:

a.   5

b.   3

c.   2

d.   7

e.   3

42.

a.   3 items at $8 each = 4 items at _____ each

b.   12 items at $3 each = 9 items at _____ each

c.   7 items at $6 each = 3 items at _____ each

d.   15 items at $10 each = 25 items at _____ each

e.   Complete the addition series: 5, 10, 15, 20, _____

SOLUTION:

a. $6

b. $4

c. $14

d. $6

e.   25

43.

$12 + 5 = 3 +$ ____

SOLUTION: 14

44.

$28 - 17 = 18 -$ ____

SOLUTION: 7

45.

$14 + 14 - 3 = 24 - 7 +$ _____

SOLUTIION: 8

46.

$120 - 99 = 7 +$ ____

SOLUTION: 14

47.

$28 + 28 = 100 -$ ____

SOLUTION: 44

48.

$14 + 7 + 3 = 9 + 7 +$ ____

SOLUTION: 8

49.

$15 \div 3 = 45 \div$ ____

SOLUTION: 9

50.

$5 \times 12 = 6 \times$ ____

SOLUTION: 10

51.

$11 \times 11 = 605 \div$ ____

SOLUTION: 5

LEARNING OBJECTIVE 5

52.
a. Complete the subtraction series: 37, 33, 29, 25, _____
b. Complete the multiplication series: 2, 4, 8, 16, _____
c. Complete the division series: 176, 88, 44, 22, _____
d. Complete the mixed series: 14, 12, 15, 13, _____
e. Complete the mixed series: 8, 32, 16, 64, _____

SOLUTION:
a. 21
b. 32
c. 11
d. 16
e. 32

53.
What is the next number in the following series: 2, 6, 7, 11?

SOLUTION: 12

54.
What is the next number in the following series: 5, 15, 45, 135?

SOLUTION: 405

55.
What is the next number if the following series: 7, 12, 10, 15, 13?

SOLUTION: 18

LEARNING OBJECTIVE 6

56.
3 tickets at $5 each = 30 tickets at _____ each

SOLUTION: $0.50

57.
7 autos at $10,000 each = 4 trucks at _____ each

SOLUTION: $17,500

58.
Compute the cost of 6 items at $2.99 each.

SOLUTION: $17.94 cost

59.
A customer purchased a pair of shoes for $39.99, a shirt for $19.99, and a belt for $9.99. All prices include tax. The customer gave the sales clerk a one hundred dollar bill. How much change did the customer receive? (Hint: Solve using rounding, then compute the exact amount.)

SOLUTION: $30.03

60.
What is the total of a purchase of 7 items at $0.99 and 5 items at $1.02?

SOLUTION: $12.03

Chapter 5   PERCENTS

PROBLEMS

LEARNING OBJECTIVES 1, 2

1.
Change the percents to decimals or mixed/whole numbers; change the decimals, fractions, or whole numbers to percents.

a.   200% = _____
b.   1 3/4 = _____
c.   0.35 = _____

SOLUTION:
a.   2

b.   175%

c.   35%

2.
Change the percents to decimals or mixed/whole numbers; change the decimals, fractions, or whole numbers to percents.

a.   1.7% = _____
b.   5/8 = _____
c.   1.7 = _____

SOLUTION:
a.   0.017

b.   62.5%

c.   170%

3.
Change the percents to decimals or mixed/whole numbers; change the decimals, fractions, or whole numbers to percents.

a.   0.27 = _____
b.   65% = _____
c.   1/2 = _____

SOLUTION:
a.   27%

b.   0.65

c.   50%

4.
Change the percents to decimals or mixed/whole numbers; change the decimals, fractions, or whole numbers to percents.

a.   0.25% = _____
b.   1/4 = _____
c.   0.125 = _____

SOLUTION:
a.   0.0025

b.   25%

c.   12.5%

5.
Solve each of the following problems for the percentage.

a.   120% of $12 = _____
b.   0.02% of 1,600 = _____
c.   40% of $250 = _____

SOLUTION:
a. $14.40   b.   0.32   c. $100

6.
Solve each of the following problems for the percentage.

a.   0.3% of 18 = _____
b.   6% of 20 = _____
c.   7.5% of $36 = _____

SOLUTION:
a.   0.054   b.   1.2   c. $2.70

LEARNING OBJECTIVE 3

7.
Solve each of the following problems for the percentage.

a.   26% of 50 = _____
b.   16% of 42.5 = _____
c.   1/4% of $220 = _____

SOLUTION:
a. 13   b.   6.8   c. $0.55

8.
Solve each of the following problems for the percentage.

a.   125% of $42.60 = _____
b.   0.325% of 100 = _____
c.   200% of $2.47 = _____

SOLUTION:
a. $53.25   b.   0.325   c. $4.94

9.
Solve each of the following for the base. Round amounts to the nearest two decimal places.

a.   25% of _____ = $4.25
b.   150% of _____ = 1.05
c.   0.375% of _____ = $3.09

SOLUTION:
a. $17   b.   0.7   c. $824

10.
Solve each of the following for the base. Round amounts to the nearest two decimal places.

a.   65% of _____ = 65
b.   1.25% of _____ = $16
c.   120% of _____ = 4.2

SOLUTION:
a.   100   b. $1,280   c.   3.5

11.
Solve each of the following for the rate. Round amounts to the nearest two decimal places.

a.   _____ of 50 = 200
b.   _____ of $316 = $553
c.   _____ of 84 = 37.8

SOLUTION:
a.   400%   b.   175%   c.   45%

12.
Solve each of the following for the rate. Round amounts to the nearest two decimal places.

a.   _____ of $15.50 = $3.72
b.   _____ of 0.15 = 0.45
c.   _____ of $400 = $10

SOLUTION:
a.   24%   b.   300%   c.   2.5%

13.
Solve each of the following for the missing value. Round amounts to the nearest two decimal places.

a.   127% of 460 = _____
b.   _____ of $220 = $9.90
c.   33 1/3% of _____ = 117

SOLUTION:
a. 584.2   b.   4.5%   c.   351

14.
Solve each of the following for the missing value. Round amounts to the nearest two decimal places.

a.   37.5% of 40,000 = _____
b.   1/5% of _____ = $4.80
c.   _____ of 2,400 = 18

SOLUTION:
a.   15,000   b. $2,400   c.   0.75%

LEARNING OBJECTIVE 4

15.
Compute the missing value.

a.   Manufacturing increased from 2,500 packages per day to 3,000 packages per day; the percent
      increased was _____
b.   Base value = $360; increase = 15%; new (final) value = _____
c.   Increasing the base value _____ by 300% gives an increase of 720 units

SOLUTION:
a.   20%   b. $414   c.   240 units

16.
Compute the missing value.

a.   An increase by 63 is 14% of the base value _____
b.   Profits were $250,000 in June, but only $220,000 in July. The rate of decrease was _____
c.   Base value = 720; decrease = 25%; new (final) value = _____

SOLUTION:
a.   450   b.   12%   c.   540

17.
Compute the missing value.

a.   The price decreased from $16.50 to $13.20; the percent decrease was _____
b.   Increasing the base value 217 by 100% yields the new value _____
c.   Decreasing the base value _____ by 75% gives a decrease of 90 units

SOLUTION:
a.   20%   b.   434   c.   120

18.
Compute the missing value.

a.   80 homes were sold in April, but only 68 homes were sold in May. The rate of decrease was
      _____
b.   Start with $120; increase by 50%; end up with _____
c.   An increase by $7.50 is 12% of the base value _____

SOLUTION:
a.   15%   b. $180   c. $62.50

LEARNING OBJECTIVE 3

19.
Nancy saves $120 per week which represents 8% of her total pay. Find the total weekly salary.

SOLUTION:
$120 ÷ 0.08 = $1,500

20.

A video store both rents and sells DVD movies and also sells blank DVD disks. During the week between Christmas and New Years, the store had rentals of $4,896, movie sales of $6,513 and blank disk sales of $2,191. What percent of the total revenue was from movie rentals?

SOLUTION:
$4,896 + $6,513 + $2,191 = $13,600; $4,896 ÷ $13,600 = 0.36 or 36%

21.

Wanda's restaurant buys all of its non-food items at a local warehouse store. On a recent trip to the warehouse store, the restaurant owner spent $126.83 just for paper products. If paper products were only 22% of the entire purchase, find the total amount spent on that trip?

SOLUTION:
$126.83 ÷ 0.22 = $576.50 total spent on shopping trip

LEARNING OBJECTIVE 4

22.

AAA Rental has 42 trailers available for rent. Last year, it had 40 available for rent. Compute the percent increase in the number of trailers.

SOLUTION:
42 − 40 = 2; 2 ÷ 40 = 0.05 = 5%

23.

Before Christmas, the price of a necktie in a Patricks Men's Shop was $42. After Christmas, the price was $25.20. By what percent was the necktie reduced?

SOLUTION:
$42 − $25.20 = $16.80; $16.80 ÷ $42 = 0.40 or 40%

LEARNING OBJECTIVE 3

24.

Darlene Davis worked a total of 45 hours, 20% of which were weekend hours. Find the number of hours that Darlene worked on the weekend.

SOLUTION:
45 hours × 0.2 = 9 hours

25.

December's electricity bill of $499.80 represents 19.6% of the total December expenses for Westcamper's Sheet Metal Fabricating. Compute Westcamper's total expenses in December.

SOLUTION:
$499.80 ÷ 0.196 = $2,550

LEARNING OBJECTIVE 4

26.

In its second year of operation, Juan's Nursery increased its business revenue by 250%. If the revenues in the second year were $175,000, what were the revenues in the first year?

SOLUTION:
$175,000 ÷ 3.5 = $50,000

27.

Karen Randolph is a marketing manager for a computer software company. When she was promoted, she received an 11% salary increase. The new monthly salary is $5,328. What was Karen's salary before the promotion?

SOLUTION:
$5,328 ÷ 1.11 = $4,800

LEARNING OBJECTIVE 3

28.

Sandy Yu earns $2,400 gross per month. Last month Sandy received a check for $1,536. What was the percent of her gross salary that had been deducted?

SOLUTION:
$2,400 – $1,536 = $864; $864 ÷ $2,400 = 0.36 = 36%

29.

Architect Brian Peters spent 60% of a week's time working on drawings for a new apartment building. If Brian spent 18 hours working on projects other than the apartment building, compute the total hours worked.

SOLUTION:
1.00 – 0.60 = 0.40; 18 hours ÷ 0.40 = 45 hours

LEARNING OBJECTIVE 4

30.
A new route for bicycling to work decreased Herschel's riding time by 12.5%. If the old route took 32 minutes, how long will the new route take?

SOLUTION:
32 minutes × 0.125 = 4 minutes; 32 minutes − 4 minutes = 28 minutes

LEARNING OBJECTIVE 3

31.
Marie Tavis is an artist. Forty percent Marie's total sales are to mail order retailers. Last year, Marie's sales to mail order retailers were $85,000. Determine her total sales last year.

SOLUTION:
$85,000 ÷ 0.40 = $212,500

32.
Pete Cassidy is an attorney for a bank. Last month, Pete worked on real estate contracts for 75% of his time. If he worked on real estate contracts for 153 hours, how long did he work on other projects?

SOLUTION:
153 hours ÷ 0.75 = 204 total hours worked
204 hours − 153 hours = 51 hours on other projects

33.
Sunrise Landscaping Company earns 85% of its annual revenue between March and October. If their total annual revenue was $254,000, how much of the total was earned either before March or after October?

SOLUTION:
100% − 85% = 15%;
$254,000 × 0.15 = $38,100 (before March or after October)

Chapter 6   COMMISSIONS

LEARNING OBJECTIVE 1

1.
Compute the commission and the total gross pay.

	Employee	Monthly Salary	Commission Rate	Monthly Sales	Commission	Gross Pay
a.	Ashley, H.	$1,800	6%	$44,500	_____	_____
b.	Bradley, R.	$1,450	6%	$38,800	_____	_____
c.	Estrada, A.	$1,050	12%	$35,000	_____	_____

SOLUTION:

a.	Commission:	$2,670;	Gross Pay:	$4,470
b.	Commission:	$2,328;	Gross Pay:	$3,778
c.	Commission:	$4,200;	Gross Pay:	$5,250

2.
Compute the commission and the total gross pay.

	Employee	Monthly Salary	Commission Rate	Monthly Sales	Commission	Gross Pay
a.	Finley, P.	$2,380	2.5%	$64,000	_____	_____
b.	McGill, F.	$1,325	6%	$64,000	_____	_____
c.	Nelson, R.	$2,142	3.75%	$28,000	_____	_____

SOLUTION:

a.	Commission:	$1,600;	Gross Pay:	$3,980
b.	Commission:	$3,840;	Gross Pay:	$5,165
c.	Commission:	$1,050;	Gross Pay:	$3,192

3.
Compute the commission and the total gross pay.

	Employee	Monthly Salary	Commission Rate	Monthly Sales	Commission	Gross Pay
a.	Pierce, D.	$1,200	4.25%	$72,360	_____	_____
b.	Quesada, T.	$0	11%	$28,200	_____	_____
c.	Sommers, P.	$3,000	1.25%	$116,800	_____	_____

SOLUTION:

a.	Commission:	$3,075.30;	Gross Pay:	$4,275.30
b.	Commission:	$3,102;	Gross Pay:	$3,102
c.	Commission:	$1,460;	Gross Pay:	$4,460

4.

Compute the commission and the total gross pay.

	Employee	Monthly Salary	Commission Rate	Monthly Sales	Commission	Gross Pay
a.	Chow, S.	$1,575	9%	$27,600	_____	_____
b.	Gomes, D.	$3,250	1.5%	$96,400	_____	_____
c.	Okutsu, K.	$2,000	4%	$41,000	_____	_____

SOLUTION:

a.	Commission:	$2,484;	Gross Pay:	$4,059
b.	Commission:	$1,446;	Gross Pay:	$4,696
c.	Commission:	$1,640;	Gross Pay:	$3,640

LEARNING OBJECTIVE 2

5.

Compute the total commission for the following commission payment plans.

	Graduated Commission Rates	Sales	Commission
a.	2% on sales to $100,000 5% on sales above $100,000	$188,500	_____
b.	1.5% on sales to $75,000 3% on sales above $75,000	$176,400	_____
c.	3% on sales to $150,000 6% on sales above $150,000	$145,000	_____

SOLUTION:

a. $188,500 - $100,000 = $88,500$; $100,000 \times 0.02 = $2,000$;
   $88,500 \times 0.05 = $4,425$; $2,000 + $4,425 = $6,425$
b. $176,400 - $75,000 = $101,400$; $75,000 \times 0.015 = $1,125$;
   $101,400 \times 0.03 = $3,042$; $1,125 + $3,042 = $4,167$
c. $145,000 \times 0.03 = $4,350$

6.

Compute the total commission for the following commission payment plans.

	Graduated Commission Rates	Sales	Commission
a.	1% on sales to $125,000 2% on sales above $125,000	$224,000	_____
b.	2.5% on sales to $65,000 3% from $65,000 to $130,000 3.5% on sales above $130,000	$195,800	_____
c.	1% on sales to $145,000 3% from $145,000 to $290,000 5% on sales above $290,000	$252,600	_____

SOLUTION:

a. $224,000 – $125,000 = $99,000; $125,000 × 0.01 = $1,250; $99,000 × 0.02 = $1,980; $1,980 + $1,250 = $3,230

b. $195,800 – $130,000 = $65,800; $65,000 × 0.025 = $1,625; $65,000 × 0.03 = $1,950; $65,800 × 0.035 = $2,303; $1,625 + $1,950 + $2,303 = $5,878

c. $252,600 – $145,000 = $91,200; $145,000 × 0.01 = $1,450; $107,600 × 0.03 = $3,228; $1,450 + $3,228 = $4,678

7.

Compute the total commission for the following commission payment plans.

	Graduated Commission Rates	Sales	Commission
a.	2% on sales to $60,000 4% from $60,000 to $120,000 6% on sales above $120,000	$147,300	_____
b.	0% on sales to $20,000 1% from $20,000 to $75,000 3% on sales above $75,000	$138,600	_____
c.	1% on sales to $75,000 2% from $75,000 to $200,000 3% on sales above $200,000	$166,200	_____

SOLUTION:

a. $147,300 – $120,000 = $27,300; $60,000 × 0.02 = $1,200; $60,000 × 0.04 = $2,400; $27,300 × 0.06 = $1,638; $1,200 + $2,400 + $1,638 = $5,238

b. $138,600 – $75,000 = $63,600; $20,000 × 0.00 = $0; $55,000 × 0.01 = $550; $63,600 × 0.03 = $1,908; $0 + $550 + $1,908 = $2,458

c. $166,200 – $75,000 = $91,200; $75,000 × 0.01 = $750; $91,200 × 0.02 = $1,824; $750 + $1,824 = $2,574

8.

Compute the total commission for the following commission payment plans.

	Graduated Commission Rates	Sales	Commission
a.	4% on sales to $25,000 5% from $25,000 to $50,000 6% on sales above $50,000	$48,880	_____
b.	3.5% on sales to $50,000 4% from $50,000 to $100,000 5.5% on sales above $100,000	$114,000	_____
c.	1% on sales to $100,000 1.5% from $100,000 to $200,000 2% on sales above $200,000	$274,000	_____

SOLUTION:

a. $48,880 – $25,000 = $23,880; $25,000 × 0.04 = $1,000;
    $23,880 × 0.05 = $1,194; $1,000 + $1,194 = $2,194

b. $114,000 – $100,000 = $14,000; $50,000 × 0.035 = $1,750;
    $50,000 × 0.04 = $2,000; $14,000 × 0.055 = $770;
    $1,750 + $2,000 + $770 = $4,520

c. $274,000 – $200,000 = $74,000; $100,000 × 0.01 = $1,000;
    $100,000 × 0.015 = $1,500; $74,000 × 0.02 = $1,480;
    $1,000 + $1,500 + $1,480 = $3,980

LEARNING OBJECTIVE 3

9.

Compute the commission and gross cost for the following purchases for principals.

	Prime Cost	Commission Rate	Trucking & Commission	Delivery	Storage	Air Freight	Gross Cost
a.	$5,350	12%	_____	$62.50	$75	$0	_____
b.	$24,200	6%	_____	$728.00	$413	$0	_____
c.	$9,650	4%	_____	$91.77	$0	$209.19	_____

SOLUTION:

a.	Commission:	$642.00;	Gross Cost:	$ 6,129.50
b.	Commission:	$1,452.00;	Gross Cost:	$26,793.00
c.	Commission:	$386.00;	Gross Cost:	$10,336.96

10.

Compute the commission and gross cost for the following purchases for principals.

	Prime Cost	Commission Rate	Trucking & Commission	Delivery	Storage	Air Freight	Gross Cost
a.	$14,150	5%	_____	$106.12	$275	$0	_____
b.	$32,500	2%	_____	$364.19	$650	$225.50	_____
c.	$26,750	8%	_____	$0	$ 98	$688.15	_____

SOLUTION:

a.	Commission:	$707.50;	Gross Cost:	$15,238.62
b.	Commission:	$650.00;	Gross Cost:	$34,389.69
c.	Commission:	$2,140.00;	Gross Cost:	$29,676.15

11.

Compute the commission and gross cost for the following purchases for principals.

	Prime Cost	Commission Rate	Trucking & Commission	Delivery	Storage	Air Freight	Gross Cost
a.	$37,450	9%	_____	$264.50	$575	$588.00	_____
b.	$14,942	8%	_____	$538.00	$224	$0	_____
c.	$21,480	11%	_____	$0	$1,050	$212.36	_____

SOLUTION:

a.	Commission:	$3,370.50;	Gross Cost:	$42,248.00
b.	Commission:	$1,195.36;	Gross Cost:	$16,899.36
c.	Commission:	$2,362.80;	Gross Cost:	$25,105.16

12.

Compute the commission and gross cost for the following purchases for principals.

	Prime Cost	Commission Rate	Trucking & Commission	Delivery	Storage	Air Freight	Gross Cost
a.	$7,750	7%	_____	$95.00	$0	$460.00	_____
b.	$6,425	6%	_____	$192.30	$175	$0	_____
c.	$9,162	5%	_____	$93.45	$0	$845.00	_____

SOLUTION:

a.	Commission:	$542.50;	Gross Cost:	$8,847.50
b.	Commission:	$385.50;	Gross Cost:	$7,177.80
c.	Commission:	$458.10;	Gross Cost:	$10,558.55

13.

Compute the commission and net proceeds for each sale on consignment.

	Gross Sales	Commission Rate	Commission	Trucking & Delivery	Storage	Air Freight	Net Proceeds
a.	$32,400	8%	_____	$175.00	$0	$381.00	_____
b.	$18,800	4.5%	_____	$808.12	$150	$0	_____
c.	$11,600	10%	_____	$0	$125	$227.75	_____

SOLUTION:

a.	Commission:	$2,592.00;	Net Proceeds:	$29,252.00
b.	Commission:	$846.00;	Net Proceeds:	$16,995.88
c.	Commission:	$1,160.00;	Net Proceeds:	$10,087.25

14.

Compute the commission and net proceeds for each sale on consignment.

	Gross Sales	Commission Rate	Commission	Trucking & Delivery	Storage	Air Freight	Net Proceeds
a.	$14,200	6%	_____	$85.75	$0	$328.00	_____
b.	$7,690	7.5%	_____	$210.00	$82	$0	_____
c.	$21,800	5%	_____	$147.50	$0	$295.00	_____

SOLUTION:

a.	Commission:	$852.00;	Net Proceeds:	$12,934.25
b.	Commission:	$576.75;	Net Proceeds:	$6,821.25
c.	Commission:	$1,090.00;	Net Proceeds:	$20,267.50

15.

Compute the commission and net proceeds for each sale on consignment.

	Gross Sales	Commission Rate	Commission	Trucking & Delivery	Storage	Air Freight	Net Proceeds
a.	$10,250	5%	_____	$219.25	$105	$0	_____
b.	$11,725	3%	_____	$107.00	$0	$412.50	_____
c.	$26,500	7%	_____	$0	$425	$535.12	_____

SOLUTION:

a.	Commission:	$512.50;	Net Proceeds:	$9,413.25
b.	Commission:	$351.75;	Net Proceeds:	$10,853.75
c.	Commission:	$1,855.00;	Net Proceeds:	$23,684.88

16.

Compute the commission and net proceeds for each sale on consignment.

	Gross Sales	Commission Rate	Commission	Trucking & Delivery	Storage	Air Freight	Net Proceeds
a.	$42,800	6%	_____	$988.00	$650	$0	_____
b.	$51,750	4%	_____	$425.00	$210	$0	_____
c.	$34,000	3%	_____	$50.00	$0	$217.20	_____

SOLUTION:
a.	Commission:	$2,568.00;	Net Proceeds:	$38,594.00
b.	Commission:	$2,070.00;	Net Proceeds:	$49,045.00
c.	Commission:	$1,020.00;	Net Proceeds:	$32,712.80

LEARNING OBJECTIVE 1

17.

An appliance salesperson receives a monthly salary of $2,250, plus a commission of 4.5% on all sales. Compute total pay for a month in which she sold $98,450 worth of appliances.

SOLUTION:
$98,450 × 0.045 = $4,430.25; $4,430.25 + $2,250 = $6,680.25

LEARNING OBJECTIVE 2

18.

A salesperson in a furniture store receives a weekly salary of $1,250. She also earns a commission of 6% on all sales for the week and a bonus of 2% on all sales above $5,000 for the week. How much would she earn if her total sales for the week were $9,420?

SOLUTION:
$9,420 − $5,000 = $4,240; $9,420 × 0.06 = $565.20;
$4,240 × 0.02 = $84.80; $565.20 + $84.80 + $1,250 = $1,900

19.

Sylvia is paid only by commission and receives 2% on all sales. If Sylvia sells more than $150,000 in any month, she receives an additional 1.5% on the sales above $150,000. Find Sylvia's total commission for August on sales of $239,000.

SOLUTION:
$239,000 − $150,000 = $89,000; $239,000 × 0.02 = $4,780;
$89,000 × 0.015 = $1,335; $4,780 + $1,335 = $6,115

LEARNING OBJECTIVE 1

20.
An employee, who sells advertising time for independent television stations, is paid a salary of $2,500 per month. For short commercials (less than 30 seconds), the employee receives no commissions. For long commercials (30 seconds or longer), the employee receives a 1/2% commission. Compute the employee's total pay for a month in which the employee sold $388,500 worth of time for short commercials, and $214,000 worth of time for long commercials.

SOLUTION:
$214,000 × 0.005 = $1,070; $1,070 + $2,500 = $3,570

LEARNING OBJECTIVE 2

21.
A free-lance sales representative is independent and gets no salary, but earns s a 5% commission on monthly sales up to $35,000, 6% on monthly sales between $35,000 and $75,000, and 7% on monthly sales over $75,000. Determine the commission for a month on sales of $106,450.

SOLUTION:
$106,450 – $75,000 = $31,450; $35,000 × 0.05 = $1,750;
$40,000 × 0.06 = $2,400.00; $31,450 × 0.07 = $2,201.50;
$1,750.00 + $2,400.00 + $2,201.50 = $6,351.50

22.
A newspaper advertisement seeking a salesperson states that pay is $1,000 salary plus 3% on sales up to $40,000 and 4% on sales over $40,000. The advertisement also states that normal sales should usually be at least $50,000. What would be the total pay for selling at the expected minimum?

SOLUTION:
$50,000 – $40,000 = $10,000; $40,000 × 0.03 = $1,200;
$10,000 × 0.04 = $400; $1,000 + $1,200 + $400 = $2,600

23.
Harley Dickerson sells industrial paper products and earns a 4% commission on all sales. Harley earns a 2% bonus on sales that exceed $40,000 during a month. An additional 1% bonus is earned on any sales that exceed $90,000. Calculate Harley's commission for a month on sales of $95,800.

SOLUTION:
$95,800 – $40,000 = $55,800; $95,800 – $90,000 = $5,800;
$95,800 × 0.04 = $3,832; $55,800 × 0.02 = $1,116;
$5,800 × 0.01 = $58; $3,832 + $1,116 + $58 = $5,006

24.

A salesperson gets paid a salary of $1,875 per month, with a graduated commission plan that pays 2% on sales up to $37,500, 2.5% on sales between $37,500 and $75,000, and 3% on sales over $75,000. Compute the commission earned on sales of $64,250.

SOLUTION:
$64,250 – $37,500 = $26,750; $37,500 × 0.02 = $750;
$26,750 × 0.025 = $668.75; $750.00 + $668.75 + $1,875.00 = $3,293.75

LEARNING OBJECTIVE 1

25.

A salesperson receives no salary, but gets an 11% commission on all sales that are final, i.e., merchandise is not returned. Last month sales were $48,500, but $7,400 worth of furnishings were returned. What was the salesperson's commission for last month?

SOLUTION:
$48,500 – $7,400 = $41,100; $41,100 × 0.11 = $4,521

26.

An employee earns a salary of $750 per week and a 4% commission on all sales. Determine the employee's total pay if her sales for the week total $18,100.

SOLUTION:
$18,100 × 0.04 = $724; $724 + $750 = $1,474

27.

Mary Wallis receives a salary of $3,750, and a 0.75% commission on all of her sales in a month. In March, she sold $260,000 worth of merchandise. Determine Mary's total pay for March.

SOLUTION:
$260,000 × 0.0075 = $1,950; $3,750 + $1,950 = $5,700

LEARNING OBJECTIVE 3

28.

Yuli Lu works as commission merchant who helps businesses sell their existing office equipment, except computers. For desks, tables, chairs and filing cabinets, Yuli charges 18% of the sales price. For printers and copiers, she charges 30%. She also deducts any delivery charges. For Harris Corp, she sold assorted furniture at $7,860 and two large copiers for $2,450 each. The total delivery charges were $318. Compute the net proceeds that would be due to Harris Corp.

SOLUTION:
$7,860 × 0.18 = $1,414.80; 2 × $2,450 = $4,900; $4,900 × 0.30 = $1,470;
$7,860 + $4,900 = $12,760; $12,760 – $1,414.80 – $1,470 – $318 = $9,557.20

29.

A commission merchant specializing in food products sold a consignment of products for a farm. The sales price was $18,450 with a commission rate of 12% for farm products. For this consignment, there was $890 freight that the merchant had to pay. What were the net proceeds due to the farm?

SOLUTION:
$18,450 × 0.12 = $2,214; $2,214 + $890 = $3,104; $18,450 − $3,104 = $15,346

30.

A commission merchant in the first quarter of the year (January through March) sold several consignments of construction equipment, some used and some new. The sales price was $112,900 for the new equipment and $58,400 for the used equipment. The merchant charges a commission rate of 10% to sell new products, and 15% to sell used equipment. The merchant paid total freight charges of $2,490 to have the equipment delivered to the various buyers. Compute the net proceeds from the sales.

SOLUTION:
$112,900 + $58,400 = $171,300; $112,900 × 0.10 = $11,290;
$58,400 × 0.15 = $8,760; $11,290 + $8,760 + $2,490 = $22,540;
$171,300 − $22,540 = $148,760

31.

An interior decorator can buy furniture at a "dealer price" but he charges his clients the "retail price." If a client spends more than $2,500 on furniture at one time, the interior decorator gives the client a 5% discount on that entire purchase. What would be the total amount of the purchase if the discount were $189.

SOLUTION:
$189 ÷ 0.05 = $3,780

32.

An landscape designer charges a modest hourly design fee. She makes the majority of her income by selling various outdoor equipment and furniture. She adds 25% to her cost and also charges any related delivery and/or "set-up" fees. A client selected furniture and barbecue equipment that cost the designer $3,274. In addition there was a delivery charge of $375 and another $247 to assemble the barbecue. Compute the total amount of the purchase.

SOLUTION:
$3,274 × 0.25 = $818.50
$3,274 + $818.50 + $375 + $247 = $4,714.50

33.

A single, working parent decides to have a home delivery service perform the weekly grocery shopping. The service charges 7.5% of the food price, plus 25 cents per mile. In one month, the service purchased food worth $586.40 and drove a total of 76 miles. Determine the total cost to the parent.

SOLUTION:
$0.25 × 76 = $19.00; $586.40 × 0.075 = $43.98;
$586.40 + $43.98 + $19.00 = $649.38

Chapter 7   DISCOUNTS

PROBLEMS

LEARNING OBJECTIVE 1

1.

Compute the amount of the trade discount and the net price using the discount method.

	List Price	Trade Discount	Discount Amount	Net Price
a.	$1,640	18%		
b.	$540	30%		
c.	$720	22%		

SOLUTION:

a.	Discount Amount:	$295.20;	Net Price:	$1,344.80
b.	Discount Amount:	$162.00;	Net Price:	$378.00
c.	Discount Amount:	$158.40;	Net Price:	$561.60

2.

Compute the amount of the trade discount and the net price using the discount method.

	List Price	Trade Discount	Discount Amount	Net Price
a.	$845	15%		
b.	$2,450	35%		
c.	$1,095	25%		

SOLUTION:

a.	Discount Amount:	$126.75;	Net Price:	$718.25
b.	Discount Amount:	$857.50;	Net Price:	$1,592.50
c.	Discount Amount:	$273.75;	Net Price:	$821.25

3.

Compute the complement rate and the net price using the complement method.

	List Price	Trade Discount	Complement Rate	Net Price
a.	$430	30%		
b.	$1,612	22%		
c.	$2,740	40%		

SOLUTION:

a.	Complement Rate:	70%;	Net Price:	$301.00
b.	Complement Rate:	78%;	Net Price:	$1,257.36
c.	Complement Rate:	60%;	Net Price:	$1,644.00

4.

Compute the complement rate and the net price using the complement method.

	List Price	Trade Discount	Complement Rate	Net Price
a.	$2,670	16%	_____	_____
b.	$904	25%	_____	_____
c.	$1,400	20%	_____	_____

SOLUTION:

a.	Complement Rate:	84%;	Net Price:	$2,242.80
b.	Complement Rate:	75%;	Net Price:	$678.00
c.	Complement Rate:	80%;	Net Price:	$1,120.00

LEARNING OBJECTIVE 2

5.

Compute the amounts of the trade discounts (where none exists, enter an "*") and the net price. Round each discount to the nearest cent.

	List Price	Trade Discounts	Trade Discount Amounts			Net Price
			First	Second	Third	
a.	$880	20%, 5%	_____	_____	_____	_____
b.	$1,680	30%, 20%, 5%	_____	_____	_____	_____
c.	$680	35%, 25%, 10%	_____	_____	_____	_____

SOLUTION:

	Trade Discount Amounts			Net Price
	First	Second	Third	
a.	$176.00	$35.20	*	$668.80
b.	$504.00	$235.20	$47.04	$893.76
c.	$238.00	$110.50	$33.15	$298.35

6.

Compute the amounts of the trade discounts (where none exists, enter an "*") and the net price. Round each discount to the nearest cent.

	List Price	Trade Discounts	Trade Discount Amounts			Net Price
			First	Second	Third	
a.	$750	35%, 20%	_____	_____	_____	_____
b.	$4,000	30%, 15%, 5%	_____	_____	_____	_____
c.	$440	25%, 15%, 10%	_____	_____	_____	_____

SOLUTION:

	Trade Discount Amounts			Net Price
	First	Second	Third	
a.	$262.50	$97.50	*	$390.00
b.	$1,200.00	$420.00	$119.00	$2,261.00
c.	$110.00	$49.50	$28.05	$252.45

7.

Compute the amounts of the trade discounts (where none exists, enter an "*") and the net price. Round each discount to the nearest cent.

	List Price	Trade Discounts	Trade Discount Amounts			Net Price
			First	Second	Third	
a.	$1,364	20%, 10%	_____			_____
b.	$6,400	20%, 15%, 10%	_____	_____	_____	_____
c.	$1,280	35%, 20%, 5%	_____	_____	_____	_____

SOLUTION:

	Trade Discount Amounts			Net Price
	First	Second	Third	
a.	$272.80	$109.12	*	$982.08
b.	$1,280.00	$768.00	$435.20	$3,916.80
c.	$448.00	$166.40	$33.28	$632.32

8.

Find the complement rates (where none exists, enter an "*") and the net price using the complement method. Do not round until the final answer.

	List Price	Trade Discounts	Complement Rates			Net Price
			First	Second	Third	
a.	$1,500	40%, 25%	_____			_____
b.	$6,400	25%, 15%, 5%	_____	_____	_____	_____
c.	$3,600	25%, 20%, 10%	_____	_____	_____	_____

SOLUTION:

	Complement Rates			Net Price
	First	Second	Third	
a.	60%	75%	*	$675.00
b.	75%	85%	95%	$3,876.00
c.	75%	80%	90%	$1,944.00

9.

Find the complement rates (where none exists, enter an "*") and the net price using the complement method. Do not round until the final answer.

	List Price	Trade Discounts	Complement Rates			Net Price
			First	Second	Third	
a.	$2,480	30%, 20%	_____			_____
b.	$2,800	25%, 15%, 10%	_____	_____	_____	_____
c.	$1,104	40%, 20%, 10%	_____	_____	_____	_____

SOLUTION:

	Complement Rates			Net Price
	First	Second	Third	
a.	70%	80%	*	$1,388.80
b.	75%	85%	90%	$1,606.50
c.	60%	80%	90%	$476.93

10.

Find the complement rates (where none exists, enter an "*") and the net price using the complement method. Do not round until the final answer.

	List Price	Trade Discounts	Complement Rates First	Second	Third	Net Price
a.	$762	20%, 10%	_____	_____	_____	_____
b.	$8,600	30%, 15%, 5%	_____	_____	_____	_____
c.	$4,250	20%, 10%, 5%	_____	_____	_____	_____

SOLUTION:

	Complement Rates First	Second	Third	Net Price
a.	80%	90%	*	$548.64
b.	70%	85%	95%	$4,861.15
c.	80%	90%	95%	$2,907.00

LEARNING OBJECTIVE 3

11.

Compute each of the complement rates and the equivalent single discount rate to the nearest 1/10 of a percent.

	Trade Discounts	Complement Rates First	Second	Third	Equivalent Single Discount
a.	35%, 15%, 5%	_____	_____	_____	_____
b.	15%, 10%, 5%	_____	_____	_____	_____
c.	30%, 25%, 15%	_____	_____	_____	_____

SOLUTION:

	Complement Rates First	Second	Third	Net Price
a.	65%	85%	95%	47.5%
b.	85%	90%	95%	27.3%
c.	70%	75%	85%	55.4%

12.

Compute each of the complement rates and the equivalent single discount rate to the nearest 1/10 of a percent.

	Trade Discounts	Complement Rates			Equivalent Single Discount
		First	Second	Third	
a.	25%, 15%, 5%	_____	_____	_____	_____
b.	35%, 25%, 10%	_____	_____	_____	_____
c.	30%, 20%, 10%	_____	_____	_____	_____

SOLUTION:

	Complement Rates			Net
	First	Second	Third	Price
a.	75%	85%	95%	39.4%
b.	65%	75%	90%	56.1%
c.	70%	80%	90%	49.6%

LEARNING OBJECTIVE 4

13.

Compute the discount date, the due date, the amount of discount and amount of the remittance if the required remittance is paid within the discount period.

Terms: 2/5, net/20		Discount Date:	_____
Invoice Date: Mar. 30		Due Date:	_____
Invoice Amount:$512.42		Discount Amount:	_____
		Remittance:	_____

SOLUTION:

Discount Date: Apr. 4
Due Date: Apr. 19
Discount Amount: $10.25
Remittance: $502.17

14.

Compute the discount date, the due date, the amount of discount and amount of the remittance if the required remittance is paid within the discount period.

Terms: 2/10, net/45		Discount Date:	_____
Invoice Date: Nov. 24		Due Date:	_____
Invoice Amount:$898.70		Discount Amount:	_____
		Remittance:	_____

SOLUTION:

Discount Date: Dec. 4
Due Date: Jan. 8
Discount Amount: $17.97
Remittance: $880.73

15.

Compute the discount date, the due date, the amount of discount and amount of the remittance if the required remittance is paid within the discount period.

Terms: 3/15, net/25   Discount Date: _____
Invoice Date: Aug. 28   Due Date: _____
Invoice Amount: $295.60   Discount Amount: _____
Returned Goods: $38.50   Remittance: _____

SOLUTION:
Discount Date: Sept. 12
Due Date: Sept. 22
Discount Amount: $7.71
Remittance: $249.39

16.

Compute the discount date, the due date, the amount of discount and amount of the remittance if the required remittance is paid within the discount period.

Terms: 2/10, net/25   Discount Date: _____
Invoice Date: Oct. 13   Due Date: _____
Invoice Amount: $819.24   Discount Amount: _____
Freight Charge: $52.12   Remittance: _____

SOLUTION:
Discount Date: Oct. 23
Due Date: Nov. 7
Discount Amount: $15.34
Remittance: $803.90

17.

Compute the discount date, the due date, the amount of discount and amount of the remittance if the required remittance is paid within the discount period.

Terms: 1/15, net/35   Discount Date: _____
Invoice Date: Dec. 27   Due Date: _____
Invoice Amount:$2,044.12   Discount Amount: _____
Freight Charge: $142.50   Remittance: _____
Returned Goods: $765.00

SOLUTION:
Discount Date: Jan. 11
Due Date: Jan. 31
Discount Amount: $11.37
Remittance: $1,267.75

18.

Compute the discount date, the due date, the amount of discount and amount of the remittance if the required remittance is paid within the discount period.

On March 29, Cathy Wellesby bought $7,420 worth of merchandise on terms of 2/10, net/45.

| Discount Date: | _____ | Discount Amount: | _____ |
| Due Date: | _____ | Remittance: | _____ |

SOLUTION:

Discount Date:	Mar. 29 + 10 days = Apr. 8
Due Date:	Mar. 29 + 45 days = May 13
Discount Amount:	$7,420 × 0.02 = $148.40
Remittance:	$7,420 − $148.40 = $7,271.60

19.

Compute the discount date, the due date, the amount of discount and amount of the remittance if the required remittance is paid within the discount period.

Sean Atchison bought various new power tools for $1,244 on March 29 on terms of 1/15, net/25. Almost immediately, he returned a power saw that cost $175.

| Discount Date: | _____ | Discount Amount: | _____ |
| Due Date: | _____ | Remittance: | _____ |

SOLUTION:

| Discount Date: | Mar. 29 + 15 days = Apr. 13 |
| Due Date: | Mar. 29 + 25 days = Apr. 23 |

$1,244 − $175 = $1,069;

Discount Amount: $1,069 × 0.01 = $10.69

Remittance: $1,069 − $10.69 = $1,058.31

20.

Compute the discount date, the due date, the amount of discount and amount of the remittance if the required remittance is paid within the discount period.

Krohn Corp. normally gives a cash discount of 2.5/5, net/20. There is no discount on freight charges. The company sells new products for which the total invoice is $142.50. The invoice is dated September 7 and includes $16.13 for freight.

| Discount Date: | _____ | Discount Amount: | _____ |
| Due Date: | _____ | Remittance: | _____ |

SOLUTION:

| Discount Date: | Sept. 7 + 5 days = Sept. 12 |
| Due Date: | Sept. 7 + 20 days = Sept. 27 |

$142.50 − $16.13 = $126.37;

| Discount Amount: | $126.37 × 0.025 = $3.16 |
| Remittance: | $126.37 − $3.16 + $16.13 = $139.34 |

21.
Compute the discount date, the due date, the amount of discount and amount of the remittance if the required remittance is paid within the discount period.

Mateski Bros. Builders Co. tries to buy building materials wherever and whenever the best cash discount is given. One supplier gives terms of 1.5/10, net/45, so Mateski Bros. bought materials which had a total invoice of $3,110. The invoice date was March 25 and included a $246 delivery charge. The builder returned some materials that cost $416. Compute the missing information.

Discount Date: _____   Discount Amount: _____

Due Date: _____   Remittance: _____

SOLUTION:

Discount Date: Mar. 25 + 10 days = Apr. 4

Due Date: Mar. 25 + 45 days = May 9

$3,110 − $246 − $416 = $2,448;

Discount Amount: $2,448 × 0.015 = $36.72

Remittance: $2,448 − $36.72 + $246 = $2,657.28

22.
Compute the discount date, the due date, the amount of discount and amount of the remittance if the entire invoice is paid within the discount period.

Jay Wilson Design offers a cash discount of 2.5/5, net/45. On June 28, Wilson Design sold goods with a total invoice of $1,724.40. Nothing was returned, but the invoice included a $90 delivery charge.

Discount Date: _____   Discount Amount: _____

Due Date: _____   Remittance: _____

SOLUTION:

Discount Date: June 28 + 5 days = July 3

Due Date: June 28 + 45 days = Aug. 12

$1,724.40 − $90 = $1,634.40

Discount Amount $1,634.40 × 0.025 = $40.86

Remittance: $1,634.60 − $40.86 + $90 = $1,683.54

23.
Compute the discount date, the due date, the amount of discount and amount of the remittance if the required remittance is paid within the discount period.

Ray Marshall is a purchasing agent for East Coast Electric and tries to take advantage of all cash discounts on major purchases. On July 7 (invoice date) he purchased parts for service vehicles. The parts cost $3,824.77, but $214.40 worth were returned because they were incorrect. The terms from the supplier were 1/10, net/25.

Discount Date: _____   Discount Amount: _____

Due Date: _____   Remittance: _____

SOLUTION:
    Discount Date: July 7 + 10 days = July 17
    Due Date: July 7 + 25 days = Aug. 1
    $3,824.77 – $214.40 = $3,610.37;
    Discount Amount: $3,610.37 × 0.01 = $36.10
    Remittance: $3,610.37 – $36.10 = $3,574.27

24.

Find the discount date, the due date, and the complement rate. Use the complement method to find the remittance if the required remittance is paid within the discount period.

Connie Chu bought some new office equipment from a dealer who offered a cash discount of 1.5/30, net/45. The invoice was dated July 20, totaled $2,652.40 and included $86 for freight.

    Discount Date: _____   Complement Rate: _____
    Due Date: _____   Remittance: _____

SOLUTION:
    Discount Date: July 20 + 30 days = Aug. 19
    Due Date: July 20 + 45 days = Sept. 3
    $2,652.40 – $86 = $2,566.40;
    Complement Rate: 100% – 1.5% = 98.5%
    $2,566.40 × 0.985 = $2,527.90
    Remittance: $2,527.90 + $86 = $2,613.90

25.

Find the discount date, the due date, and the complement rate. Use the complement method to find the remittance if the required remittance is paid within the discount period.

Evelyn Haynes Brown is a management consultant who recommended to a client that the client try offering cash discount terms of 0.5/15, net/25. The first invoice prepared under the new plan was for $890 and was dated November 1. Of the total, $48 was for freight and another $214 worth of merchandise was quickly returned.

    Discount Date: _____   Complement Rate: _____
    Due Date: _____   Remittance: _____

SOLUTION:
    Discount Date: Nov. 1 + 15 days = Nov. 16
    Due Date: Nov. 1 + 25 days = Nov. 26
    $890 – $48 – $214 = $628;
    Complement Rate: 100% – 0.5% = 99.5%
    $628 × 0.995 = $624.86
    Remittance: $624.86 + $48 = $672.86

26.

Find the discount date, the due date, and the complement rate. Use the complement method to find the remittance if the required remittance is paid within the discount period.

Lakewood Resorts buys canned food on terms of 1/5, net/45. An invoice for canned food dated July 3 was for $374. The total included a $24 delivery charge. On July 5, Lakeside Resorts returned cans worth $74.20 because they were damaged.

    Discount Date: _____   Complement Rate: _____
    Due Date: _____   Remittance: _____

SOLUTION:

    Discount Date: July 3 + 5 days = July 8
    Due Date: July 3 + 45 days = Aug. 17
    $374 - $24 - $74.20 = $275.80;
    Complement Rate: 100% - 1% = 99%
    $275.80 × 0.99 = $273.04
    Remittance: $273.04 + $24 = $297.04

27.

Find the discount date, the due date, and the complement rate. Use the complement method to find the remittance if the required remittance is paid within the discount period.

Oceanview Market buys salami and other cured meats on terms of 2.5/10, net/30. An invoice for cured meats was dated July 26 and was for $697.16. On July 28 the deli returned $128.60 worth of the cured meats because they were not packaged properly.

    Discount Date: _____   Complement Rate: _____
    Due Date: _____   Remittance: _____

SOLUTION:

    Discount Date: July 26 + 10 days = Aug. 5
    Due Date: July 26 + 30 days = Aug. 25
    $697.16 - $128.60 = $568.56;
    Complement Rate: 100% - 2.5% = 97.5%
    Remittance: $568.56 × 0.975 = $554.35

28.

Find the discount date, the due date, and the complement rate. Use the complement method to find the remittance if the required remittance is paid within the discount period.

A new supply company offered cash discount terms of 3/5, net/30. On January 24, a new customer purchased items with an invoice total of $738.40. Then, two days later, the customer returned $346.40 worth of the items.

    Discount Date: _____   Complement Rate: _____
    Due Date: _____   Remittance: _____

SOLUTION:
Discount Date: Jan. 24 + 5 days = Jan. 29
Due Date: Jan. 24 + 30 days = Feb. 23
$738.40 – $346.40 = $392.00;
Complement Rate: 100% – 3% = 97%
Remittance: $392 × 0.97 = $380.24

29.

Find the discount date, the due date, and the complement rate. Use the complement method to find the remittance if the required remittance is paid within the discount period.

In an attempt to attract new business customers, the wholesale hardware store offered terms of 2.5/40, net/60. On July 10, a building contractor purchased materials costing $611.25, which included a $50 freight charge. On August 2, the contractor returned materials costing $143.25.

Discount Date: _____   Complement Rate: _____
Due Date: _____   Remittance: _____

SOLUTION:
Discount Date: July 10 + 40 days = Aug. 19
Due Date: July 10 + 60 days = Sept. 8
$611.25 – $50 – $143.25 = $418.00;
Complement Rate: 100% – 2.5% = 97.5%
Remittance: $418 × 0.975 = $407.55; $407.55 + $50 = $457.55

LEARNING OBJECTIVE 5

30.

Contractor Jason Barrett bought $8,750 worth of building materials on August 26. The supplier offered a 2.5% cash discount on any amount paid within 7 days. Compute the amount the builder should remit to credit their account for $7,500. (This problem involves the partial payment of an invoice within the discount period.)

SOLUTION:
100% – 2.5% = 97.5%
$7,500 × 0.975 = $7,312.50 remittance

31.

Brianna Carson purchased some tools with a total cost of $442.52. The invoice was dated February 27 and the seller gave terms of 2/10, net/30. The discount applies to any amount paid within the discount period. Compute the amount that Brianna should remit to credit her account for $300. (This problem involves the partial payment of an invoice within the discount period.)

SOLUTION:
100% – 2% = 98%
$300 × 0.98 = $294 remittance

32.
Tony Chang gives good customers a cash discount of 1.5% on any amount that they pay within 5 days of the invoice date. An invoice for $1,890 was written on February 2. The customer paid $900 on February 6. Compute the amount that Tony must credit to his customer's account. (This problem involves the partial payment of an invoice within the discount period.)

SOLUTION:
100% − 1.5% = 98.5%
$900 ÷ 0.985 = $913.71 credited

33.
Susan Waterford bought merchandise for $1,880.14 on March 3. There was a 3% cash discount on anything paid within 5 days. Susan sent in a check for $1,200 on March 5. Compute the amount that should credited to Susan's account. (This problem involves the partial payment of an invoice within the discount period.)

SOLUTION:
100% − 3% = 97%
$1,200 ÷ 0.97 = $1,237.11 credited

34.
Winthrop Racing owns a car racing team on the East Coast. On May 25, Winthrop purchased $26,750 worth of racing tires for the cars in its team. To get the Winthrop account, the tire manufacturer offered generous terms of 5/15, net/60. The discount applies to any amount paid within the discount period. What is the unpaid balance if Winthrop paid $20,000 on June 5? (This problem involves the partial payment of an invoice within the discount period.)

SOLUTION:
100% − 5% = 95%
$20,000 ÷ 0.95 = $21,052.63
$26,750 − $21,052.63 = $5,697.37 unpaid balance

35.
Seven years ago, Elise Damato started a Doggy Day Care franchise business. After seven years, her pet food supplier gives Elise a cash discount of 2.75% on any amount paid within 12 days of the invoice date. On August 29, Elise bought pet food costing $1,245.60. On September 10, Elise paid $1,200 on the purchase. Compute the unpaid balance. (This problem involves the partial payment of an invoice within the discount period.)

SOLUTION:
100% − 2.75% = 97.25%
$1,200 ÷ 0.9725 = $1,233.93
$1,245.60 − $1,233.93 = $11.67 unpaid balance

36.

On October 19, Kindler Welding Corp. purchased $545 worth of welding supplies from a wholesaler that offers terms of 2.5/10, net/25. The discount applies to any amount paid within the discount period. What is the unpaid balance if the welding shop paid $400 on October 25? (This problem involves the partial payment of an invoice within the discount period.)

SOLUTION:
100% – 2.5% = 97.5%
$400 ÷ 0.975 = $410.26
$545 – $410.26 = $134.74 unpaid balance

37.

Donna Pierce manages a business that distributes imported woolen fabrics. They offer a 2% cash discount on any amount paid within 15 days of the invoice date. On June 20, a purchase for fabrics priced at $1,650 was made and on July 3 a $1,000 check was sent in. Compute the unpaid balance. (This problem involves the partial payment of an invoice within the discount period.)

SOLUTION:
100% – 2% = 98%
$1,000 ÷ 0.98 = $1,020.41
$1,650 – $1,020.41 = $629.59 unpaid balance

Chapter 8   MARKUP

PROBLEMS

LEARNING OBJECTIVE 1

1.
Compute the missing terms.

	Cost	Dollar Markup	Selling Price
a.	$307.12	_____	$549.99
b.	92.50	85.00	_____
c.	_____	182.40	395.50

SOLUTION:
a. $549.99 – $307.12 = $242.87 dollar markup
b. $92.50 + $85.00 = $177.50 selling price
c. $395.50 – $182.40 = $213.10 cost

2.
Compute the missing terms.

	Cost	Dollar Markup	Selling Price
a.	_____	$275.50	$847.77
b.	105.24	94.25	_____
c.	1,100.00	_____	1,950.00

SOLUTION:
a. $847.77 – $275.50 = $572.27 cost
b. $105.24 + $94.25 = $199.49 selling price
c. $1,950 – $1,100 = $850 dollar markup

3.
Compute the missing terms.

	Cost	Dollar Markup	Selling Price
a.	$290.25	_____	$739.97
b.	_____	638.20	1,350.00
c.	490.00	510.00	_____

SOLUTION:
a. $739.97 – $290.25 = $449.72 dollar markup
b. $1,350.00 – $638.20 = $711.80 cost
c. $490 + $510 = $1,000 selling price

4.

Compute the missing terms.

	Cost	Dollar Markup	Selling Price
a.	$725.50	$424.50	_____
b.	_____	819.70	1,450.00
c.	446.61	_____	825.49

SOLUTION:

a. $725.50 + $424.50 = $1,150 selling price
b. $1,450.00 – $819.70 = $630.30 cost
c. $825.49 – $446.61 = $378.88 dollar markup

LEARNING OBJECTIVE 2

5.

The markup percent is based on cost. Compute the missing terms.

	Cost	Markup Percent	Dollar Markup	Selling Price
a.	$2,240	125%	_____	_____
b.	$825	30%	_____	_____
c.	1,200	65%	_____	_____

SOLUTION:

a.   125% × $2,240 = $2,800 dollar markup;
     $2,240 + $2,800 = $5,040 selling price
b.   30% × $825 = $247.50 dollar markup;
     $825 + $247.50 = $1,072.50 selling price
c.   65% × $1,200 = $780 dollar markup; $1,200 + $780 = $1,980 selling price

6.

The markup percent is based on cost. Compute the missing terms.

	Cost	Markup Percent	Dollar Markup	Selling Price
a.	416	75%	_____	_____
b.	$1,175	150%	_____	_____
c.	212	225%	_____	_____

SOLUTION:

a.   75% × $416 = $312 dollar markup; $416 + $312 = $728 selling price
b.   150% × $1,175 = $1,762.50 dollar markup;
     $1,175 + $1,762.50 = $2,937.50 selling price
c.   225% × $212 = $477 dollar markup; $212 + $477 = $689 selling price

7.

The markup percent is based on cost. Compute the missing terms.

	Cost	Markup Percent	100% + Markup %	Selling Price
a.	$62	25%	_____	_____
b.	880	75%	_____	_____
c.	1,640	80%	_____	_____

SOLUTION:
a.   100% + 25% = 125%; 125% × $62 = $77.50 selling price
b.   100% + 75% = 175%; 175% × $880 = $1,540 selling price
c.   100% + 80% = 180%; 180% × $1,640 = $2,952 selling price

8.

The markup percent is based on cost. Compute the missing terms.

	Cost	Markup Percent	100% + Markup %	Selling Price
a.	$1,088	75%	_____	_____
b.	475	120%	_____	_____
c.	1,500	200%	_____	_____

SOLUTION:
a.   100% + 75% = 175%; 175% × $1,088 = $1,904 selling price
b.   100% + 120% = 220%; 220% × $475 = $1,045 selling price
c.   100% + 200% = 300%; 300% × $1,500 = $4,500 selling price

LEARNING OBJECTIVES 2, 3

9.

The markup percent is based on cost. Compute the missing terms.

	Cost	Markup Percent	100% + Markup %	Selling Price
a.	_____	80%	_____	$81.90
b.	_____	100%	_____	114.00
c.	_____	140%	_____	2,148.00

SOLUTION:
a.   100% + 80% = 180%; $81.90 ÷ 180% = $45.50 cost
b.   100% + 100% = 200%; 114 ÷ 200% = $57 cost
c.   100% + 140% = 240%; 2,148 ÷ 240% = $895 cost

10.

The markup percent is based on cost. Compute the missing terms.

	Cost	Markup Percent	100% + Markup %	Selling Price
a.	_____	25%	_____	$1,555.75
b.	_____	60%	_____	78.40
a.	_____	200%	_____	417.00

SOLUTION:

a.   100% + 25% = 125%; $1,555.75 ÷ 125% = $1,244.60 cost
b.   100% + 60% = 160%; 78.40 ÷ 160% = $49 cost
a.   100% + 200% = 300%; 417 ÷ 300% = $139 cost

11.

The markup percent is based on cost. Compute the missing terms.

	Cost	Markup Percent	Dollar Markup	Selling Price
a.	$960	_____	_____	$1,248
b.	225	_____	_____	495
c.	340	_____	_____	850

SOLUTION:

a. $1,248 – $960 = $288 dollar markup;
      $288 ÷ $960 = 0.3 = 30% markup percent based on cost
b. $495 – $225 = $270 dollar markup;
      $270 ÷ $225 = 1.20 = 120% markup percent based on cost
c. $850 – $340 = $510 dollar markup;
      $510 ÷ $340 = 1.5 = 150% markup percent based on cost

12.

The markup percent is based on cost. Compute the missing terms.

	Cost	Markup Percent	Dollar Markup	Selling Price
a.	$1,060	_____	_____	$3,180
b.	3,500	_____	_____	6,300
c.	80	_____	_____	100

SOLUTION:

a. $3,180 – $1,060 = $2,120 dollar markup;
      $2,120 ÷ $1,060 = 2.00 = 200% markup percent based on cost
b. $6,300 – $3,500 = $2,800 dollar markup;
      $2,800 ÷ $3,500 = 0.80 = 80% markup percent based on cost
c. $100 – $80 = $20 dollar markup;
      $20 ÷ $80 = 0.25 = 25% markup percent based on cost

LEARNING OBJECTIVE 4

13.

Markup percent is based on selling price. Compute the missing terms.

	Selling Price	Markup Percent	100% − Markup	Cost
a.	$2,514.50	50%	_____	_____
b.	518.00	60%	_____	_____
c.	4,500.00	15%	_____	_____

SOLUTIION:
a. 100% − 50% = 50%; 50% × $2,514.50 = $1,257.25 cost
b. 100% − 60% = 40%; 40% × $518 = $207.20 cost
c. 100% − 15% = 85% × $4,500 = $3,825 cost

14.

Markup percent is based on selling price. Compute the missing terms.

	Selling Price	Markup Percent	100% − Markup	Cost
a.	$1,080.00	40%	_____	_____
b.	24.50	10%	_____	_____
c.	72.00	35%	_____	_____

SOLUTION:
a. 100% − 40% = 60%; 60% × $1,080 = $648 cost
b. 100% − 10% = 90%; 90% × $24.50 = $22.05 cost
c. 100% − 35% = 65%; 65% × $72 = $46.80 cost

15.

Markup percent is based on selling price. Compute the missing terms.

	Selling Price	Markup Percent	Dollar Markup	Cost
a.	$898.00	25%	_____	_____
b.	2,624.00	50%	_____	_____
c.	96.00	30%	_____	_____

SOLUTION:
a. $898 × 25% = $224.50 dollar markup; $898 − $224.50 = $673.50 cost
b. $2,624 × 50% = $1,312 dollar markup; $2,624 − $1,312 = $1,312 cost
c. $96 × 30% = $28.80 dollar markup; $96 − $28.80 = $67.20 cost

16.
Markup percent is based on selling price. Compute the missing terms.

	Selling Price	Markup Percent	Dollar Markup	Cost
a.	$450.75	60%	_____	_____
b.	35.00	80%	_____	_____
c.	824.00	75%	_____	_____

SOLUTION:
a. $450.75 × 60% = $270.45 dollar markup; $450.75 – $270.45 = $180.30 cost
b. $35 × 80% = $28 dollar markup; $35 – $28 = $7 cost
c. $824 × 75% = $618 dollar markup; $824 – $618 = $206 cost

LEARNING OBJECTIVES 4, 5

17.
The markup percent is based on selling price. Compute the missing terms.

	Selling Price	Markup Percent	100% – Markup	Cost
b.	_____	90%	_____	$100
a.	_____	65%	_____	3,115
c.	_____	20%	_____	296

SOLUTION:
b.   100% – 90% = 10%; $100 ÷ 10% = $1,000 selling price
a.   100% – 65% = 35%; $3,115 ÷ 35% = $8,900 selling price
c.   100% – 20% = 80%; $296 ÷ 80% = $370 selling price

18.
The markup percent is based on selling price. Compute the missing terms.

	Selling Price	Markup Percent	100% – Markup	Cost
b.	_____	15%	_____	$2,431
a.	_____	80%	_____	568
c.	_____	30%	_____	1,491

SOLUTION:
b.   100% – 15% = 85%; $2,431 ÷ 85% = $2,860 selling price
a.   100% – 80% = 20%; $568 ÷ 20% = $2,840 selling price
c.   100% – 30% = 70%; $1,491 ÷ 70% = $2,130 selling price

19.
The markup percent is based on selling price. Compute the missing terms.

	Selling Price	Markup Percent	Dollar Markup	Cost
a.	$2,420	_____	_____	$968
b.	195	_____	_____	78
c.	35	_____	_____	21

SOLUTION:
a. $2,420 – $968 = $1,452 dollar markup;
      $1,452 ÷ $2,420 = 0.60 = 60% markup percent based on selling price
b. $195 – $78 = $117 dollar markup;
      $117 ÷ $195 = 0.60 = 60% markup percent based on selling price
c. $35 – $21 = $14 dollar markup;
      $14 ÷ $35 = 0.40 = 40% markup percent based on selling price

20.
The markup percent is based on selling price. Compute the missing terms.

	Selling Price	Markup Percent	Dollar Markup	Cost
a.	$1,100	_____	_____	$770
b.	1,700	_____	_____	1,105
c.	908	_____	_____	681

SOLUTION:
a. $1,100 – $770 = $330 dollar markup;
      $330 ÷ $1,100 = 0.30 = 30% markup percent based on selling price
b. $1,700 – $1,105 = $595 dollar markup;
      $595 ÷ $1,700 = 0.35 = 35% markup percent based on selling price
c. $908 – $681 = $227 dollar markup;
      $227 ÷ $908 = 0.25 = 25% markup percent based on selling price

LEARNING OBJECTIVE 1

21.
J.D. Straley owns New England Art Gallery and specializes in selling water color paintings of the New England landscape. Although he usually sells on consignment, recently Straley bought a small painting from a well-known artist for $3,450 and marked it up by $2,725. Compute the selling price of the painting.

SOLUTION:
$3,450 + $2,725 = $6,175 selling price

22.

Bill Lounsberry's hobby was repairing old foreign sports cars in a garage at home. To store parts, Lounsberry bought some steel industrial warehouse shelving. Each unit was 6 feet long, 7 feet tall, and 2 feet deep with 500 pound capacity for each shelf. The distributor sold each unit for $425. Calculate the cost per unit to the distributor if the dollar markup was $92.30.

SOLUTION:
$425.00 − $92.30 = $332.70 cost

23.

Jon Draper sold a small, lightweight garden tractor to a landscape company for use on landscaping jobs in small areas. The tractor had a cost of $3,190 and Draper sold it for $4,795. Compute the dollar markup.

SOLUTION:
$4,795 − $3,190 = $1,605 dollar markup

24.

Because of possible dangerous fumes, a new state health law required the owners of cleaners and dry cleaners to install large air circulators (fans). Gretchen Farmer, who sells such equipment, has a large rotating air circulator on a 5-ft weighted stand that cost $242.20. Gretchen marked it up by $161.29. Find the selling price of the air circulator.

SOLUTION:
$242.20 + $161.29 = $403.49 selling price

25.

Adrienne Schweitzer manages an office supply store. A combination printer/copier/fax machine recently sold for $395. Adrienne knows that the fax machine cost the store $209. How much was the dollar markup?

SOLUTION:
$395 − $209 = $186 dollar markup

26.

Jasmine Chandler recently opened a new office as a Certified Public Accountant (CPS). Jasmine bought a fireproof four-drawer vertical filing cabinet for $857.40. If the vendor had marked it up by $229, how much did the vendor pay for the filing cabinet?

SOLUTION:
$857.40 − $229 = $628.40 cost

LEARNING OBJECTIVE 3

27.

David Tang operates an automotive wrecking yard and he needs a 4-ton chain hoist to move heavy car and truck parts. Tang can buy a used hoist for $3,840 from a dealer who paid $1,600 for it. Compute the markup percent based on cost.

SOLUTION:
$3,840 − $1,600 = $2,240 dollar markup;
$2,240 ÷ $1,600 = 1.4 or 140% markup percent based on cost

LEARNING OBJECTIVE 2

28.
Michael Waxman, a restaurant consultant, advised a client to buy a 30-inch diameter commercial kitchen ventilator which was specifically designed to remove "grease-laden" air. Waxman knew that most vendors marked up these items 75% based on cost. If the cost of such a ventilator, with motor, is $1,560, what is the selling price?

SOLUTION:
100% + 75% = 175%; 175% × $1,560 = $2,730 selling price
Alternate Solution: 75% × $1,560 = $1,170 dollar markup;
                          $1,560 + $1,170 = $2,730 selling price

29.
Dana Josephson manages a marina with her sister, Darlene. To moor sailboats to the dock, the sisters needed rope. A store was selling 5/8 inch twisted strand polypropylene rope in 600 foot rolls. The store had paid $123.64 per roll for the rope. For the sisters, the hardware store marked up the rope 20% above their cost. Compute the dollar markup and the selling price that the sisters would pay for one roll of the rope.

SOLUTION:
20% × $123.64 = $24.73 dollar markup;
$123.64 + $24.73 = $148.37 selling price

LEARNING OBJECTIVE 3

30.
Hagen Banse is a landscape contractor. He charges less for labor than other contractors, but he charges his clients the retail price for a plant while he pays only a wholesale price. Hagen charges a client $450 for an olive tree that only costs him $250. What is the markup percent based on cost?

SOLUTION:
$450 − $250 = $200 dollar markup
$200 ÷ $250 = 0.8 or 80% markup percent based on cost

31.
Sonny Fujito sells machine shop equipment. One customer was interested in a 4-speed, metal-cutting, horizontal band saw, without motor. This particular band saw cost $2,464. If Sonny sold the saw for $4,004, what was the percent markup percent based on cost?

SOLUTION:
$4,004 − $2,464 = $1,540 dollar markup;
$1,540 ÷ $2,464 = 0.625 = 62.5% based on cost

LEARNING OBJECTIVE 2

32.
Won Yin imports furniture and sells through outlet stores. Yin's markup for imported tables is 125% based on cost. If a table sells for $877.50, how much did it cost?

SOLUTION:
100% + 125% = 225%; $877.50 ÷ 225% = $390.00 cost

33.
Julianne Fong started a company which sells equipment to retrofit buildings for the physically challenged. Julianne will pay $485.60 for a wheel chair access water cooler, with front and side push bars to activate the water fountain. If she decides to mark up the price 37.5% based on cost, what will be the selling price of the water cooler?

SOLUTION:
100% + 37.5% = 137.5%; 137.5% × $485.60 = $667.70 selling price
Alternate Solution: 37.5% × $485.60 = $182.10 dollar markup;
                $485.60 + $182.10 = $667.70 selling price

LEARNING OBJECTIVE 5

34.
Omar Davies owns a large sporting goods store with several departments, one of which sells only camping equipment. Omar buys one model of portable gas stove for $60. During the camping season, Omar sells the stove for $187.50. What is the markup percent based on selling price?

SOLUTION:
$187.50 – $60 = $127.50 dollar markup
$127.50 ÷ $187.50 = 0.68 or 68% markup percent based on selling price

LEARNING OBJECTIVE 2

35.
Dorothy Quinones manages a security equipment store that sells items including alarm systems, lock and safes. Dorothy marks up everything in the store 50% based on cost. If she sells a 5,040 cubic inch · fireproof safe for $474.51, compute the dollar markup on the safe.

SOLUTION:
100% + 50% = 150%; $474.51 ÷ 150% = $316.34 cost;
$474.51 – $316.34 = $158.17 dollar markup

LEARNING OBJECTIVE 3

36.
Michela Sosa sells roofing supplies and equipment to contractors. Michela marks up all equipment items based on cost. Her selling price on a pneumatic roofing stapler is $540. If the cost of the stapler had been $297, what is the markup percent based on selling price?

SOLUTION:
$540 – $297 = $243 dollar markup;
$243 ÷ $540 = 0.45= 45% markup percent based on selling price

LEARNING OBJECTIVE 5

37.
Harold Danielson sold kitchen equipment to homeowners although he made his store appear as if it were a restaurant supply store. Harold was selling a meat slicing machine for $480 and it had cost him $168. What was the percent markup percent based on selling price?

SOLUTION:
$480 – $168 = $312 dollar markup;
$312 ÷ $480 = 0.65 or 65% markup percent based on selling price

LEARNING OBJECTIVE 4

38.
Kevin Woo owns a small shop which specializes in selling scaffolding equipment, ladders, and painting supplies. Kevin marks up all ladders 30% based on selling price. A 28-foot aluminum extension ladder has a cost of $196.14. What is the dollar markup?

SOLUTION:
100% – 3% = 70%; $196.14 ÷ 70% = $280.20 selling price;
$280.20 – $196.14 = $84.06 dollar markup

39.
Wilson Gomez manages the large screen television sales for a high-volume, discount appliance dealer. Large screen televisions are normally marked up only 40% based on selling price. Compute the dollar markup and the dealer's cost on a large screen television set that sells for $2,190.

SOLUTION:
40% × $2,190 = $876 markup; $2,190 – $876 = $1,314 cost

LEARNING OBJECTIVE 3

40.
Near the city park, Wendy Haskell opened a bicycle shop where she rents and sells both mountain and racing bikes. Wendy buys her rental mountain bikes from a local manufacturer who makes the bikes at a cost of $48 each. If Wendy pays $78 for each bike, what is the manufacturer's markup percent based on cost?

SOLUTION:
$78 – $48 = $30 dollar markup;
$30 ÷ $48 = 0.625 of 62.5% markup percent based on cost

LEARNING OBJECTIVE 5

41.
Cesare Aguilar works as a salesperson for an office products warehouse store. The markup percent on laser printers is based on selling price. One popular, competitively priced laser printer sells for $520. Cesare knows that the cost of this item is $312. Compute the markup percent based on selling price.

SOLUTION:
$520 – $312 = $208 dollar markup;
$208 ÷ $520 = 0.40 = 40% markup percent based on selling price

LEARNING OBJECTIVE 4

42.
Janelle Austin services hot tubs and spas and sells replacement parts. Janelle marks up replacement parts 30% based on selling price. Compute the selling price of a hot tub water heater that has a cost of $107.17.

SOLUTION:
100% – 30% = 70%; $107.17 ÷ 70% = $153.10 selling price

LEARNING OBJECTIVE 5

43.
As a source of income after retirement, Ian Donnelly sells an assortment of hard-to-find, specialty hardware items out of his home. Ian buys a pair of small "ratchet-style" pruning clippers for $15 each and sells them for $45 each. What is the markup percent based on selling price?

SOLUTION:
$45 – $15 = $30 dollar markup;
$30 ÷ $45 = 0.667 or 66.7% markup percent based on selling price

LEARNING OBJECTIVE 4

44.
Steve Sather sells plumbing equipment and supplies. Sather will mark up a 50-gallon residential hot water heater 40% based on selling price. If the hot water heater costs $228.96, what will be the selling price?

SOLUTION:
100% – 40% = 60%; $228.96 ÷ 60% = $381.60 selling price

45.
Brad Pham works as a department manager in a computer products store. One particular fax machine sells for $212.40. Management told Pham that the markup on this fax machine was 45% based on selling price. Calculate the cost of the machine.

SOLUTION:
100% – 45% = 55%; 55% × $212.40 = $116.82 cost
Alternate Solution: 45% × $212.40 = $95.58 dollar markup;
                    $212.40 – $95.58 = $116.82 cost

LEARNING OBJECTIVE 5

46.
Neale Burgraaf wants to purchase a halogen lamp for her office. Neal found one that is selling for $250. Through a friend, Neal learns that the dealer's cost of this lamp is $105.00. Find the dealer's markup percent based on selling price.

SOLUTION:
$250 – $105 = $145 dollar markup;
$145 ÷ $250 = 58% markup percent based on selling price

Chapter 9   BANKING

LEARNING OBJECTIVE 1

1.
A First National Bank depositor made out a deposit slip showing currency of $610.00, coins of $13.25, and two checks for $68.00 and $171.00. Compute the total deposit shown on the deposit slip.

SOLUTION: $862.25

2.
A Bank of the West deposit slip shows currency of $970.00, three checks listed on the front for $58.92, $37.46, and $4.59, and total checks from the other side of the deposit slip of $687.40. Compute the total shown on the deposit slip.

SOLUTION: $1,758.37

3.
A deposit slip from Commercial Bank of Oakland shows a total deposit of $1,078.42. The cash received line shows $70.00. Compute the net deposit shown on the deposit slip.

SOLUTION: $1,008.42

4.
A check register shows the current cash balance to be $1,260.40. A check was written after this cash balance date for $193.84. Compute the new cash balance shown in the check register.

SOLUTION: $1,066.56

5.
The check register for Tennet Company shows a balance of $981.40 before adding a deposit of $22.90 and subtracting checks of $200.00 and $17.38. Compute the new cash balance shown in the check register.

SOLUTION: $786.92

6.
A check register shows a balance of $391.63 before adding a deposit of $12.33 and writing three checks for $3.78, $4.92, and $11.17. Compute the new cash balance shown in the check register.

SOLUTION: $384.09

7.
A check register for the Stephen Clark Wholesale Company showed a balance of $1,103.21 before two checks for $935.62 and $11.11 were written and before a deposit of $3,704.28 was made. Compute the new cash balance shown in the check register.

SOLUTION: $3,860.76

8.

The check register for Promotions, Inc. showed a balance of $5,684.37. The bookkeeper then made a deposit of $6,753.91. The bookkeeper then wrote checks for $1,704.12, $1,600.32, and $2,110.14. Compute the new cash balance shown in the check register.

SOLUTION: $ 7,023.70

9.

A check register shows a balance of $787.77 before adding a deposit of $110.75 and writing two checks for $117.50 and $605.00. Compute the new cash balance shown in the check register.

SOLUTION: $176.02

10.

A check register shows a balance of $147.20 before writing three checks for $63.24, $31.54, and $86.50. Compute the new cash balance shown in the check register.

SOLUTION: $34.08

11.

The checkbook stubs for Marcia Ayala showed a balance of $303.95 before adding two deposits of $125.17 and $369.18. She then wrote checks for $117.82 and $99.99. Compute the new cash balance shown on the check stub after the check for $99.99 was written.

SOLUTION: $580.49

12.

A check register shows a balance of $3.75 before adding two deposits of $0.91 and $210.35, and writing two checks for $147.98 and $39.46. Compute the new cash balance shown in the check register.

SOLUTION: $27.57

13.

The check register for Bellam Furniture showed a balance of $3,511.91. Checks were written for $2,593.02 and $72.80. A deposit was made for $6,982.22. Compute the new balance shown in the check register.

SOLUTION: $7,828.31

14.

A check register shows a balance of $18.23 before adding deposits of $491.04, $52.65, and $22.33. Compute the new cash balance shown in the check register.

SOLUTION: $584.25

15.

A check register shows a balance of $900.00 before adding a deposit of $111.11 and writing two checks for $921.34 and $89.77. Compute the new cash balance shown in the check register.

SOLUTION: $0.00

16.
A check register shows a balance of $861.64 before writing four checks for $31.11, $157.03, $71.44, and $92.38. Compute the new cash balance shown in the check register.

SOLUTION: $509.68

17.
A check register shows a balance of $19.20 before adding six deposits of $3.97, $1.84, $5.25, $28.14, $28.15 and $28.16. Compute the new cash balance shown in the check register.

SOLUTION: $114.71

18.
The check stub in Alan Nelson's checkbook showed a balance of $675.00 before Alan wrote two checks for $464.00 and $101.00, and made a deposit of $100.00. Compute the new cash balance.

SOLUTION: $210.00

19.
A check register shows a balance of $661.10 before writing checks for $13.42, $21.78, $1.08, $0.57, and $444.44. Compute the new cash balance shown in the check register.

SOLUTION: $179.81

20.
A check register shows a balance of $191.83 before adding deposits of $18.19, $31.00, $65.25, $10.46, and $23.64. Compute the new cash balance shown in the check register.

SOLUTION: $340.37

21.
A check register shows a balance of $398.42 before writing two checks for $73.50 and $225.00. Compute the amount the account owner must deposit before writing a third check for the amount of $110.00.

SOLUTION: $10.08

22.
A check register shows a balance of $845.50 before adding a deposit of $152.73 and writing three checks for $682.67, $146.05, and $99.33. The owner of the checking account needs to write two additional checks for $75.81 and $83.62. Compute the amount the account owner must deposit before writing the additional checks.

SOLUTION: $89.25

23.
A check register shows a balance of $171.11 before adding deposits of $400.00 and $421.87. Three checks for $22.75, $618.90, and $35.00 are then written. Compute the amount of the new cash balance shown in the check register.

SOLUTION: $316.33

24.

A bank statement shows a balance of $347.72. The check register of the account owner shows outstanding checks of $119.35, $75.00, and $150.00. Compute the adjusted cash balance of the bank statement.

SOLUTION: $3.37

25.

A check register shows a balance of $146.73. The bank statement shows a service charge of $6.50, a charge of $3.00 for a photocopy of a check as requested by the account owner, and interest of $2.37 credited to the account. Compute the adjusted cash balance of the check register.

SOLUTION: $139.60

26.

A check register shows a balance of $1,180.94. The bank statement shows that interest of $5.61 was credited by the bank, and a charge of $11.50 was made by the bank for printing new checks. Compute the adjusted cash balance of the check register.

SOLUTION: $1,175.05

27.

A bank statement shows a balance of $2,970.31 The check register of the account owner shows an outstanding deposit of $1,482.93 and outstanding checks of $2,500.00, $1,000.00, and $500.00. Compute the adjusted cash balance of the bank statement.

SOLUTION: $453.24

28.

A check register shows a balance of $152.34. The bank statement shows that a check for $75.00 deposited by the account owner was drawn against insufficient funds and was returned. A charge for $2.00 was also deducted by the bank because of the return. Compute the adjusted cash balance of the check register.

SOLUTION: $75.34

29.

A check register shows a balance of $247.82. The bank statement shows a service charge of $7.50. The bank statement also shows that the bank paid interest of $1.21. Compute the adjusted cash balance of the check register.

SOLUTION: $241.53

30.

A bank statement shows a balance of $1.85. The check register of the account owner shows outstanding deposits of $350.00 and $247.32, as well as an outstanding check of $475.50. Compute the adjusted cash balance of the bank statement.

SOLUTION: $123.67

31.

A check register shows a balance of $346.71. The bank statement shows that two checks of $42.34 and $56.25, deposited by the account owner, were drawn against a closed account and were returned as invalid. A charge for $2.00 for each deposit was also deducted by the bank because of the returns. Compute the adjusted cash balance of the check register.

SOLUTION: $244.12

32.

A check register shows a balance of $3,740.38. The bank statement shows that the bank paid interest of $12.37 received an electronic deposit for $2,000.00 and charged $19.00 for an NSF returned check. Compute the adjusted cash balance of the check register.

SOLUTION: $5,733.75

33.

A bank statement shows a balance of $471.84. The check register of the account owner shows outstanding checks of $117.73, $50.00, and $14.46 and an outstanding deposit of $4,320.01. Compute the adjusted cash balance of the bank statement.

SOLUTION: $4,609.66

34.

A check register shows a balance of $260.50. The bank statement shows a service charge of $7.00. The bank statement also shows interest of $1.34 credited to the account. Compute the adjusted cash balance of the check register.

SOLUTION: $254.84

35.

A check register shows a balance of $995.57. The bank statement shows a charge of $10 for a photocopy of a check as requested by the account owner, a regular service charge of $6.50, and a special charge of $12.00 for the printing of personalized checks. The bank statement also shows that interest of $1.49 was paid by the bank. Compute the adjusted cash balance of the check register.

SOLUTION: $968.56

36.

A bank statement shows a balance of $198.37. The check register of the account owner shows outstanding deposits of $12.70, $134.65, and $190.00. Compute the adjusted cash balance of the bank statement.

SOLUTION: $535.72

37.

A check register shows a balance of $170.35. The bank statement shows interest of $1.42 credited to the account and a service charge of $6.50 charged against the account. Compute the adjusted cash balance of the check register.

SOLUTION: $165.27

38.

A bank statement shows a balance of $78,620.14. The check register shows an outstanding deposit of $11,007.21 and outstanding checks of $180.90, $2,384.12 and $17,000.50. Compute the adjusted cash balance of the bank statement.

SOLUTION: $70,061.83

39.

A check register shows a balance of $347.52. The bank statement shows a cash balance of $356.02. The check register shows an outstanding check for $15.00. The bank statement shows a service charge of $6.50. Compute the adjusted cash balance of the check register and the bank statement.

SOLUTION: $341.02

40.

A check register shows a balance of $484.95. The bank statement shows a cash balance of $506.27. The check register shows an outstanding check for $27.82. The bank statement shows a service charge of $6.50. Compute the adjusted balance of the check register and the bank statement.

SOLUTION: $478.45

41.

A check register shows a balance of $383.75. The bank statement shows a cash balance of $285.82. The check register shows an outstanding deposit of $142.93 and two outstanding checks of $37.41 and $12.65. The bank statement shows a service charge of $7.00 and an interest deposit of $1.94. Compute the adjusted cash balance of the check register and the bank statement.

SOLUTION: $378.69

42.

A check register shows a balance of $671.94. The bank statement shows a cash balance of $632.75. The check register shows an outstanding deposit of $183.97 and two outstanding checks of $42.36 and $105.10. The bank statement shows a service charge of $7.00 and an interest deposit of $4.32. Compute the adjusted cash balance of the check register and the bank statement.

SOLUTION: $669.26

43.

A check register shows a balance of $1,062.55. The bank statement shows a cash balance of $1,156.35. The check register shows outstanding deposits of $93.72 and $59.84 as well as outstanding checks of $31.71, $58.32, and $168.23. The bank statement shows a service charge of $7.00, a charge of $10.00 for a photocopy of a check, and an interest deposit of $6.10. Compute the adjusted cash balance of the check register and the bank statement.

SOLUTION: $1,051.65

44.

A check register shows a balance of $1,447.36. The bank statement shows a cash balance of $1,572.53. The check register shows an outstanding deposit of $1,000.00 and two outstanding checks of $750.00 and $390.00. The bank statement shows a service charge of $6.75, a charge of $12.00 for printing checks, and an interest deposit of $3.92. Compute the adjusted cash balance of the check register and the bank statement.

SOLUTION: $1,432.53

45.

A check register shows a balance of $3,683.52. The bank statement shows a cash balance of $10,020.18. The check register shows two outstanding checks of $1,348.31 and $5,000.00. The bank statement shows a service charge of $11.65. Compute the adjusted cash balance of the check register and the bank statement.

SOLUTION: $3,671.87

LEARNING OBJECTIVE 2

46.

Compute the reconciled balance:
Bank statement balance; $869.55
Checkbook balance; $914.78
Outstanding checks; $43.27 and $29.50
Automatic transfer to savings; $100.00
Automatic charge, safe deposit box; $18.00

SOLUTION: $796.78

47.

Compute the reconciled balance:
Bank statement balance; $642.75
Checkbook balance; $685.23
Outstanding checks; $14.82 and $117.70
Outstanding deposit; $125.00
Automatic transfer to savings; $50.00

SOLUTION: $635.23

48.

Compute the reconciled balance:
Bank statement balance; $375.59
Checkbook balance; $462.17
Outstanding checks; $3.50 and $25.00
Automatic transfer to savings; $100.00
Automatic charge, safe deposit box; $12.00
Service charge; $6.50
Bank interest credited; $3.42

SOLUTION: $347.09

49.
Compute the reconciled balance:
Bank statement balance; $261.47
Checkbook balance; $335.12
Outstanding deposits; $108.50 and $213.17
Outstanding checks; $18.25, $22.50, and $325.00
Service charge; $9.40
Automatic charge for printing checks; $11.00
Automatic transfer to savings; $100.00
Bank interest credited; $2.67

SOLUTION: $217.39

50.
Compute the reconciled balance:
Bank statement balance; $137,614.28
Checkbook balance; $_____
Outstanding deposits; $23,618.20 and $31,046.12
Outstanding checks; $17,249.17, $52,076.14, and $997.38

SOLUTION: $121,955.91

51.
Indicate whether each of the following should be (A) added to the bank statement balance, (B) subtracted from the bank statement balance, (C) added to the checkbook balance, or (D) subtracted from the checkbook when computing the reconciled balance.

a.   Deposit in transit to the bank
b.   Bank service charge
c.   Outstanding checks
d.   Interest on the account
e.   Error in checkbook: A check for $55 was recorded as $35

SOLUTION: A, D, B, C, D

52.
Indicate whether each of the following should be (A) added to the bank statement balance, (B) subtracted from the bank statement balance, (C) added to the checkbook balance, or (D) subtracted from the checkbook when computing the reconciled balance.

a.   Bank error: A deposit of $280 is shown on the statement as $250
b.   NSF check returned to the depositor
c.   Automatic transfer for safe deposit fee
d.   Error in the checkbook: a check written for $45 was recorded as $54
e.   ATM withdrawal not recorded in the depositor's records

SOLUTION: A, D, D, C, D

53.
Indicate whether each of the following should be (A) added to the bank statement balance, (B) subtracted from the bank statement balance, (C) added to the checkbook balance, or (D) subtracted from the checkbook when computing the reconciled balance.

a. Outstanding checks
b. Bank fee for printing checks
c. Error in the check register: a deposit of $995 was recorded as $955
d. Deposit which did not appear on the bank statement
e. Service fees for ATM withdrawals

SOLUTION: B, D, C, A, D

54.
A check stub shows a balance brought forward of $935.65, a deposit of $416.20, and a check written for $116.89. Calculate the balance carried forward.

SOLUTION: $1,234.96

55.
A check stub shows a balance brought forward of $1,620.34, a deposit of $825.45, and a check amount of $369.22. Calculate the balance carried forward.

SOLUTION: $2,076.57

56.
A check stub shows a balance brought forward of $835.90, a deposit of $250.65, and a check amount of $85.62. Calculate the balance carried forward.

SOLUTION: $1,000.93

57.
The check register of Boswell Co. indicated a cash balance of $3,689.50 at the beginning of the month. During the month Boswell Co. made two deposits of $1,450.34 and $950.30. During the same month Boswell Co. wrote checks for $140.50, $1,469.29, $51.90, and $789.12. What balance should the register show for cash at the end of the month?

SOLUTION: $3,639.33

58.
The check register of Clover Co. indicated a cash balance of $1,580.12 January 1. During the month Clover Co. made deposits of $980.75 and $2,120.85. During the same month Clover Co. wrote checks for $140.50, $1,469.29, $690.17, $51.90, and $789.12. What balance should the register show for cash January 31?

SOLUTION: $1,540.74

59.

Calculate the reconciled cash balance:

Bank statement balance	$2,835.18
Checkbook balance	2,416.80
Deposits in transit	820.12
Outstanding checks	1,250.50
Service fees	12.00

SOLUTION: $2,404.80

60.

Calculate the reconciled cash balance:

Bank statement balance	$5,602.50
Checkbook balance	6,105.06
Deposits in transit	1,355.25
Outstanding checks	872.19
Bank service charges	19.50

SOLUTION: $6,085.56

Chapter 10 PAYROLL RECORDS

PROBLEMS

LEARNING OBJECTIVE 1

1.
The Kelly Services Company pays the regular hourly rate for the first 40 hours worked and time and a half thereafter. George Wilson worked 45 hours this week. His regular hourly rate is $12.00. Compute Wilson's total earnings.

SOLUTION: $570

2.
Mossdale Manufacturing Company pays the regular hourly rate for the first 40 hours worked and time and a half thereafter. Sandra Beal worked 46 hours this week. Her regular hourly rate is $15.60. Compute Beal's total earnings.

SOLUTION: $764.44

3.
The Burton Company pays the regular hourly rate for the first 40 hours worked and time and a half thereafter. Adolph Mercer worked 51 hours this week. His regular hourly rate is $16.12. Compute Mercer's total earnings.

SOLUTION: $910.78

4.
The Greenwood Company pays the regular hourly rate for the first 36 hours worked and time and a half thereafter. Dolly Webb worked 40 hours this week. Her regular hourly rate is $10.50. Compute Webb's total earnings.

SOLUTION: $441.00

5.
The Zapata Motor Works pays the regular hourly rate for the first 40 hours worked and time and a half thereafter. John Simpson worked 54 hours this week. His regular hourly rate is $8.50. Compute Simpson's total earnings.

SOLUTION: $518.50

6.
Cal Transit Company pays the regular hourly rate for the first 40 hours worked, time and a half for hours 41 through 48, and double time for all hours in excess of 48. Gayle Jordon worked 50 hours this week. Her regular hourly rate is $10.50. Compute Jordan's total earnings.

SOLUTION: $588

7.

The Mason Company pays the regular hourly rate for the first 40 hours worked, time and a half for hours 41 through 48, and double time for all hours in excess of 48. Frank Davis worked 47 hours this week. His regular hourly rate is $8.40. Compute Davis's total earnings.

SOLUTION: $424.20

8.

Decor Paints pays regular hourly rates for the first 40 hours worked and time and a half thereafter. The weekly payroll record shown below gives the total hours worked and hourly rates for the three employees. Fill in the remaining information including totals.

Payroll Record
DECOR PAINTS
Weekly Payroll
Week Ending June 4, 20____

Name	Total Hours	Regular Earnings Hours Worked	Regular Earnings Rate/ Hour	Regular Earnings Amount	Overtime Earnings Hours Worked	Overtime Earnings Rate/ Hour	Overtime Earnings Amount	Total Earnings
Allen, R.	45	_____	$7.40	_____	_____	_____	_____	_____
Brady, J.	52	_____	$7.10	_____	_____	_____	_____	_____
Cole, D.	48	_____	$8.60	_____	_____	_____	_____	_____
Totals				_____			_____	_____

SOLUTION:

Payroll Record
DECOR PAINTS
Weekly Payroll
Week Ending June 4, 20____

Name	Total Hours	Regular Earnings Hours Worked	Regular Earnings Rate/ Hour	Regular Earnings Amount	Overtime Earnings Hours Worked	Overtime Earnings Rate/ Hour	Overtime Earnings Amount	Total Earnings
Allen, R.	45	40	$7.40	$296	5	$11.10	$ 55.50	$ 351.50
Brady, J.	52	40	$7.10	$284	12	$10.65	$127.80	$ 411.80
Cole, D.	48	40	$8.60	$344	8	$12.90	$103.20	$ 447.20
Totals				$924			$286.50	$1,210.50

Allen:	$7.40 \times 40 = 296; $7.40 \times 1.5 = 11.10 overtime rate
	$45 - 40 = 5$; $5 \times $11.10 = 55.50; $296 + $55.50 = 351.50 total
Brady:	$7.10 \times 40 = 284; $7.10 \times 1.5 = 10.65 overtime rate
	$52 - 40 = 12$; $12 \times $10.65 = 127.80; $284 + $127.80 = 411.80 total
Cole:	$8.60 \times 40 = 344; $8.60 \times 1.5 = 12.90 overtime rate
	$48 - 40 = 8$; $8 \times $12.90 = 103.20; $344 + 103.20 = 447.20 total
Totals:	$296 + $284 + $344 = 924 regular
	$55.50 + $127.80 + $103.2 = 286.50 overtime earnings
	$924 + $286.50 = $1,210.50$ total earnings

9.

Seaside Patio Furniture pays regular hourly rates for the first 40 hours worked and time and a half thereafter. The weekly payroll record shown below gives the total hours worked and hourly rates for the three employees. Fill in the remaining information including totals.

Payroll Record
SEASIDE PATIO FURNITURE
Weekly Payroll
Week Ending May 12, 20___

		Regular Earnings			Overtime Earnings			
Name	Total Hours	Hours Worked	Rate/ Hour	Amount	Hours Worked	Rate/ Hour	Amount	Total Earnings
Garcia, R.	42	_____	$12.90	_____	_____	_____	_____	_____
Kelly, J.	38	_____	$10.25	_____	_____	_____	_____	_____
Obi, D.	48	_____	$14.70	_____	_____	_____	_____	_____
Totals				_____			_____	_____

SOLUTION:

Payroll Record
SEASIDE PATIO FURNITURE
Weekly Payroll
Week Ending May 12, 20___

		Regular Earnings			Overtime Earnings			
Name	Total Hours	Hours Worked	Rate/ Hour	Amount	Hours Worked	Rate/ Hour	Amount	Total Earnings
Garcia, R.	42	40	$12.90	$516	2	$19.35	$38.70	$554.70
Kelly, J.	38	38	$10.25	$399	0	$15.75	0	$399.00
Obi, D.	48	40	$14.70	$588	8	$22.05	$176.40	$764.50
Totals				$1,503			$215.10	$1,718.10

LEARNING OBJECTIVE 3

10.

The required deduction for social security is 6.2% OASDI (Old Age Survivors and Disability Insurance) of wages to a maximum of $87,900 and 1.45% HI (Hospital Insurance) for all wages. Sonia Evans earns a monthly salary of $2,800. Compute the amount deducted each month for social security.

SOLUTION: $214.20

11.

The required deduction for social security is 6.2% OASDI (Old Age Survivors and Disability Insurance) of wages earned, to a maximum of $87,900 and 1.45% HI (Hospital Insurance) for all earnings. Enrique Marsh earns a monthly salary of $4,200. Compute the amount deducted over the period of a year for social security.

SOLUTION: $3,855.60

12.

The required deduction for social security is 6.2% OASDI (Old Age Survivors and Disability Insurance) of wages earned, to a maximum of $87,900 and 1.45% HI (Hospital Insurance) for all earnings. Compute the deduction for social security during the period of a year assuming a person earns $90,000.

SOLUTION: $6,754.8   .062 × 87900 = 5449.8 + .0145 × 90000 = 1305

13.

The required deduction for social security if 6.2% OASDI (Old Age Survivors and Disability Insurance) of wages earned, to a maximum of $87,900 and 1.45% HI (Hospital Insurance) for all earnings. The employer is then required to match the employee's deduction and send the total to the IRS. Compute the maximum percent that can be sent to the IRS for any one employee during the period of a year.

SOLUTION: 15.3%

14.

The required deduction for social security is 6.2% OASDI (Old Age Survivors and Disability Insurance) of wages earned, to a maximum of $87,900 and 1.45% HI (Hospital Insurance) for all earnings. The employer must match the employee's deduction and send the total to the IRS. Compute the amount of money that would be sent to the IRS for an employee who earns $52,700 during the period of a year.

SOLUTION: $8,063.10

15.

The required deduction for social security is 6.2% OASDI (Old Age Survivors and Disability Insurance) of wages earned, to a maximum of $87,900 and 1.45% HI (Hospital Insurance) for all earnings. Keri Albertson earns a monthly salary of $8,500. Compute the amount deducted for social security from Albertson's January paycheck.

SOLUTION: $650.25

16.

The required deduction for social security is 6.2% OASDI (Old Age Survivors and Disability Insurance) of wages earned, to a maximum of $87,900 and 1.45 HI (Hospital Insurance) for all earnings. Keri Albertson earns a monthly salary of $8,500. Compute the amount deducted for social security from Albertson's November paycheck.

SOLUTION: 303.05

17.

The required deduction for social security is 6.2% OASDI (Old Age Survivors and Disability Insurance) of wages earned, to a maximum of $87,900 and 1.45% HI (Hospital Insurance) for all earnings. Bill Smith earns a monthly salary of $9,000. Compute the amount deducted for social security from Smith's November paycheck.

SOLUTION: $130.50

18.

The required deduction for social security is 6.2% OASDI (Old Age Survivors and Disability Insurance) of wages earned, to a maximum of $87,900 in earnings. Beverly McCoy earns a monthly salary of $4,950. Compute the amount deducted for social security from McCoy's December paycheck.

SOLUTION: $378.68

19.

The required deduction for social security is 6.2% OASDI (Old Age Survivors and Disability Insurance) of wages earned, to a maximum of $87,900 and 1.45% HI (Hospital Insurance) for all earnings. The employer must match the employee's deduction and send the total to the IRS. An employer has 15 employees, each earning $27,000 per year; 5 employees, each earning $32,000 a year; and 2 employees, each earning $45,000 a year. Compute the amount the employer pays in matching social security funds during the year.

SOLUTION: $50,107.50

20.

The required deduction for social security is 6.2% OASDI (Old Age Survivors and Disability Insurance) of wages earned, to a maximum of $87,900 and 1.45% HI (Hospital Insurance) for all earnings. Marge Brown's cumulative wages for the year not including the current pay period total $86,500. Calculate the amount of social security to be withheld from Brown's current gross pay of $2,800.

SOLUTION: $127.40

21.

The required deduction for social security is 6.2% OASDI (Old Age Survivors and Disability Insurance) of wages earned, to a maximum of $87,900 and 1.45% HI (Hospital Insurance) for all earnings. Mable Jay's cumulative wages for the year not including the current pay period total $92,100. Calculate the amount of social security to be withheld from Jay's current gross pay of $3,825.

SOLUTION: $55.46

LEARNING OBJECTIVE 2

22.

A withholding allowance for the employee and each dependent is exempt from gross earnings and not subject to federal income tax. The monthly deduction for this allowance is $258.33. Laura Wilder claims as a dependent herself only. Compute the total amount of Wilder's allowance.

SOLUTION: $258.33

23.

A withholding allowance for the employee and each dependent is exempt from gross earnings and not subject to federal income tax. The monthly deduction for this allowance is $258.33. Chris Kent claims himself and one dependent. Compute the total amount of Kent's allowance.

SOLUTION: $516.66

24.

A withholding allowance for the employee and each dependent is exempt from gross earnings and not subject to federal income tax. The monthly deduction for this allowance is $258.33. Darla Olson claims herself and two dependents. Compute the total amount of Olson's allowance.

SOLUTION: $774.99

25.

A withholding allowance for the employee and each dependent is exempt from gross earnings and not subject to federal income tax. The monthly deduction for this allowance is $258.33. Jake Rogers claims himself and three dependents. Compute the total amount of Rogers' allowance.

SOLUTION: $1,033.32

26.

A withholding allowance for the employee and each dependent is exempt from gross earnings and not subject to federal income tax. The monthly deduction for this allowance is $258.33. Clara Webb claims herself and four dependents. Compute the total amount of Webb's allowance.

SOLUTION: $1,291.65

27.

A withholding allowance for the employee and each dependent is exempt from gross earnings and not subject to federal income tax. The monthly deduction for this allowance is $258.33. Tim Brown claims himself and three dependents. His monthly salary is $2,000. Compute Brown's taxable income per month.

SOLUTION: $966.68

28.

A withholding allowance for the employee and each dependent is exempt from gross earnings and not subject to federal income tax. The monthly deduction for this allowance is $258.33. Janis Perry claims herself and two dependents. Her monthly salary is $2,800. Compute Perry's taxable income per month.

SOLUTION: $2,025.01

29.
A withholding allowance for the employee and each dependent is exempt from gross earnings and not subject to federal income tax. The monthly deduction for this allowance is $258.33. Mark Fox claims himself and four dependents. His monthly salary is $3,600. Compute Fox's taxable income per month.

SOLUTION: $2,308.35

30.
A withholding allowance for the employee and each dependent is exempt from gross earnings and not subject to federal income tax. The monthly deduction for this allowance is $258.33. Mary Farnsworth, who has a monthly salary of $2,600, claims herself and one dependent. Compute Farnsworth's taxable income per month.

SOLUTION: $2,083.34

31.
A withholding allowance for the employee and each dependent is exempt from gross earnings and not subject to federal income tax. The monthly deduction for this allowance is $258.33. James Webb, who has a monthly salary of $2,450, claims himself and three dependents. Compute Webb's taxable income per month.

SOLUTION: $1,416.68

32.
A withholding allowance for the employee and each dependent is exempt from gross earnings and not subject to federal income tax. The monthly deduction for this allowance is $258.33. Kitty Wilson, has a monthly salary of $2,820, and claims only herself. Compute Wilson's taxable income per month.

SOLUTION: $2,561.67

33.
A withholding allowance for the employee and each dependent is exempt from gross earnings and not subject to federal income tax. The monthly deduction for this allowance is $258.33. Jennifer Souza is a student who works part time. Since she is claimed on her parents' return as a dependent, she claims zero exemptions. How much of Souza's monthly wages of $750 is taxable?

SOLUTION: $750

LEARNING OBJECTIVE 1

34.
A weekly payroll register shows that Louise Price had total wages of $460. She had deductions of $35.19 for FICA tax, $54 for federal income tax, and $16 for medical insurance. Compute Price's net pay.

SOLUTION: $354.81

35.
A weekly payroll register shows that James Hill had total wages of $380. He had deductions of $29.07 for FICA tax, $42.50 for federal income tax, and $20 for medical insurance. Compute Hill's net pay.

SOLUTION: $288.43

36.

A weekly payroll register shows that Phillip Kelly had total wages of $451.25. He had deductions of $34.52 for FICA tax, $57.20 for federal income tax, and $36 for medical insurance. Compute Kelly's net pay.

SOLUTION: $323.53

37.

A weekly payroll register shows that Virginia Peters had total wages of $520. She had deductions of $39.78 for FICA tax, $64.50 for federal income tax, and $17 for medical insurance. Compute Peters' net pay.

SOLUTION: $398.72

38.

A weekly payroll register shows that Teresa Miller had total wages of $402. She had deductions of $30.75 for FICA tax, $43.50 for federal income tax, and $22 for medical insurance. Compute Miller's net pay.

SOLUTION: $305.75

39.

A weekly payroll register shows that Will Morris had total wages of $390. He had deductions of $29.83 for FICA tax, $41.20 for federal income tax, and $17.50 for medical insurance. Compute Morris' net pay.

SOLUTION: $301.47

40.

A weekly payroll register shows that William Phillips had total wages of $381. He had deductions of $29.15 for FICA tax, $39.70 for federal income tax, and $24 for medical insurance. Compute Phillips' net pay.

SOLUTION: $288.15

41.

A weekly payroll register shows that Carla Mitchell had total wages of $402. She had deductions of $30.75 for FICA tax, $46.40 for federal income tax, and $18 for medical insurance. Compute Mitchell's net pay.

SOLUTION: $306.85

42.

A monthly payroll register shows that Sean Dougan had total wages of $3,580. He had deductions of $273.87 for FICA tax, $335.50 for federal income tax, and $25 for medical insurance. Compute Dougan's net pay for the month.

SOLUTION: $2,945.63

43.

A monthly payroll register shows that Andrew Li had total wages of $2,845. He had deductions of $217.64 for FICA tax, $198 for federal income tax, and $40 for medical insurance. Compute Dougan's net pay for the month.

SOLUTION: $2,389.36

LEARNING OBJECTIVE 2

44.

Herb Becker had gross earnings of $465 last week. He is single and claims a withholding allowance for himself only. The amount of each weekly withholding allowance is $59.62. Using the table for single taxpayers below, compute the amount that will be withheld from his paycheck for federal income tax. (Figure withholding to the nearest cent.)

Over	But not over	
$51	$187	10% of excess over $51
$187	$592	$13.60 plus 15% of excess over $187
$592	$1,317	$74.35 plus 25% of excess over $592
$1,317	$2,860	$255.60 plus 28% of excess over $1,317
$2,860	$6.177	$687.64 plus 33% of excess over $2,860
$6,177		$1,782.25 plus 35% of excess over $6,177

SOLUTION: $46.36

45.

Jay Murray had gross earnings of $250 last week. He is single and claims only himself as an exemption. The weekly withholding allowance for each exemption is $59.62. Using the table for single taxpayers below, figure the amount that will be withheld from his paycheck for federal income tax. (Figure withholding to the nearest cent.)

Over	But not over	
$51	$187	10% of excess over $51
$187	$592	$13.60 plus 15% of excess over $187
$592	$1,317	$74.35 plus 25% of excess over $592
$1,317	$2,860	$255.60 plus 28% of excess over $1,317
$2,860	$6.177	$687.64 plus 33% of excess over $2,860
$6,177		$1,782.25 plus 35% of excess over $6,177

SOLUTION: $ 14.11 $190.38 - 187 = 3.38 \times .15 = .51 + 13.6 = 14.11$

46.

Rose White had gross earnings of $545 last week. She is single and claims an allowance for herself only. The amount of one weekly withholding allowance is $59.62. Using the table for single taxpayers below, figure the amount that will be withheld from her paycheck for federal income tax. (Figure withholding to the nearest cent.)

Over	But not over	
$51	$187	10% of excess over $51
$187	$592	$13.60 plus 15% of excess over $187
$592	$1,317	$74.35 plus 25% of excess over $592
$1,317	$2,860	$255.60 plus 28% of excess over $1,317
$2,860	$6.177	$687.64 plus 33% of excess over $2,860
$6,177		$1,782.25 plus 35% of excess over $6,177

SOLUTION: $58.36

47.

Amy McGreggor had gross earnings of $740 last week. She is single and claims an allowance for herself only. The amount of one weekly withholding allowance is $59.62. Using the table for single taxpayers below, figure the amount that will be withheld from her paycheck for federal income tax. (Figure withholding to the nearest cent.)

Over	But not over	
$51	$187	10% of excess over $51
$187	$592	$13.60 plus 15% of excess over $187
$592	$1,317	$74.35 plus 25% of excess over $592
$1,317	$2,860	$255.60 plus 28% of excess over $1,317
$2,860	$6.177	$687.64 plus 33% of excess over $2,860
$6,177		$1,782.25 plus 35% of excess over $6,177

SOLUTION: $96.45

48.

John Davis had gross earnings of $500 last week. He is single and claims an allowance for himself only. The amount of one weekly withholding allowance is $59.62. Using the table for single taxpayers below, figure the amount that will be withheld from her paycheck for federal income tax. (Figure withholding to the nearest cent.)

Over	But not over	
$51	$187	10% of excess over $51
$187	$592	$13.60 plus 15% of excess over $187
$592	$1,317	$74.35 plus 25% of excess over $592
$1,317	$2,860	$255.60 plus 28% of excess over $1,317
$2,860	$6.177	$687.64 plus 33% of excess over $2,860
$6,177		$1,782.25 plus 35% of excess over $6,177

SOLUTION: $51.61

49.

Earl Denny had gross earnings of $980 last week. He is married and claims allowances for himself, his wife, and two dependent children. The weekly withholding allowance for each dependent is $59.62. Using the table for married taxpayers below, figure the amount that will be withheld from his paycheck for federal income tax. (Figure withholding to the nearest cent.)

Over	But not over	
$154	$429	10% of excess over $154
$429	$1,245	$27.50 plus 15% of excess over $429
$1,245	$2,270	$149.90 plus 25% of excess over $1,245
$2,270	$3,568	$406.15 plus 28% of excess over $2,270
$3,568	$6,271	$769,59 plus 33% of excess over $3,568
$6,271		$1,661.58 plus 35% of excess over $6,271

SOLUTION: $74.38

50.

Carl Smith had gross earnings of $850 last week. He is married and claims himself, his wife, and three children as dependents. The weekly withholding allowance for each dependent is $59.62. Using the table for married taxpayers below, figure the amount that will be withheld from his paycheck for federal income tax. (Figure withholding to the nearest cent.)

Over	But not over	
$154	$429	10% of excess over $154
$429	$1,245	$27.50 plus 15% of excess over $429
$1,245	$2,270	$149.90 plus 25% of excess over $1,245
$2,270	$3,568	$406.15 plus 28% of excess over $2,270
$3,568	$6,271	$769,59 plus 33% of excess over $3,568
$6,271		$1,661.58 plus 35% of excess over $6,271

SOLUTION: $45.94

51.

Douglas Johnson had gross earnings of $2,350 last week. He is married and claims himself and his wife as dependents. The weekly withholding allowance for each dependent is $59.62. Using the table for married taxpayers below, figure the amount that will be withheld from his paycheck for federal income tax. (Figure withholding to the nearest cent.)

Over	But not over	
$154	$429	10% of excess over $154
$429	$1,245	$27.50 plus 15% of excess over $429
$1,245	$2,270	$149.90 plus 25% of excess over $1,245
$2,270	$3,568	$406.15 plus 28% of excess over $2,270
$3,568	$6,271	$769,59 plus 33% of excess over $3,568
$6,271		$1,661.58 plus 35% of excess over $6,271

SOLUTION: $396.34

52.

Susan Smith had gross earnings of $2,800 last week. She is married and claims two allowances on her W-4. The amount of each weekly withholding allowance is $59.62. Using the table for married taxpayers below, figure the amount that will be withheld from her paycheck for federal income tax. (Figure withholding to the nearest cent.)

Over	But not over	
$154	$429	10% of excess over $154
$429	$1,245	$27.50 plus 15% of excess over $429
$1,245	$2,270	$149.90 plus 25% of excess over $1,245
$2,270	$3,568	$406.15 plus 28% of excess over $2,270
$3,568	$6,271	$769,59 plus 33% of excess over $3,568
$6,271		$1,661.58 plus 35% of excess over $6,271

SOLUTION: $521.16

53.

Zachery Thompson had gross earnings of $540 last week. He is married and claims two allowances on his W-4. The amount of each weekly withholding allowance is $59.62. Using the table for married taxpayers below, figure the amount that will be withheld from his paycheck for federal income tax. (Figure withholding to the nearest cent.)

Over	But not over	
$154	$429	10% of excess over $154
$429	$1,245	$27.50 plus 15% of excess over $429
$1,245	$2,270	$149.90 plus 25% of excess over $1,245
$2,270	$3,568	$406.15 plus 28% of excess over $2,270
$3,568	$6,271	$769,59 plus 33% of excess over $3,568
$6,271		$1,661.58 plus 35% of excess over $6,271

SOLUTION: $26.68

LEARNING OBJECTIVE 4

54.

Jay Murray's earnings record shows thirteen weeks of consistent earnings and deductions for the first quarter as follows: Weekly wage $480, Federal Income Tax withholding $57, Social Security withholding $29.76, Medicare withholding $6.96, Group medical insurance deductions $12. Compute the totals which would appear on his employee's earnings record for the quarter.

SOLUTION:

Total Wages	$6,240.00
Federal Income Tax withholding	741.00
Social Security withholding	386.88
Medicare withholding	90.48
Group medical insurance deductions	156.00
Total Deductions	$1,374.36
Net Pay	$4,865.64

55.

Rose White's earning record shows thirteen weeks of consistent earnings and deductions for the first quarter as follows: Weekly wage $545, Federal Income tax withholding $66.31, Social Security withholding $33.85, Medicare withholding $7.90, Group dental plan deductions $7.50. Compute the totals which would appear on her employee's earnings record for the quarter.

SOLUTION:

Total Wages	$7,085.00
Federal Income Tax withholding	862.03
Social Security withholding	440.05
Medicare withholding	102.70
Group dental plan deductions	97.50
Total Deductions	$1,502.28
Net Pay	$5,582.72

LEARNING OBJECTIVE 5

56.

For the first quarter of 2002, Denton Toys paid total wages of $175,000.38. The company withheld $26,500 for federal income tax. All wages paid were subject to Social Security and Medicare taxes. If during the quarter Denton Toys had deposited $50,000 toward its taxes due, how much would be required to send in with its first-quarter Form 941?

SOLUTION:

Gross wages $175,000.38 × 12.4% Social Security	$21,700.05
Gross wages $175,000.38 × 2.9% Medicare	5,075.01
Subtotal	$26,775.06
Income taxes withheld	26,500.00
Total	$53,275.06
Less deposit	50,000.00
Balance	$ 3,275.06

57.

For the first quarter of 2002, Tover Paint Company paid total wages of $400,000. The company withheld $58,000 for federal income tax. All wages paid were subject to Social Security and Medicare taxes. If during the quarter Tover Paint Company had deposited $115,000 toward its taxes due, how much would be required to send in with its first-quarter Form 941?

SOLUTION:

Gross wages $400,000 × 12.4% Social Security	$49,600.00
Gross wages $400,000 × 2.9% Medicare	11,600.00
Subtotal	$61,200.00
Income taxes withheld	58,000.00
Total	$119,200.00
Less deposit	115,000.00
Balance	$4,200.00

LEARNING OBJECTIVE 6

58.

During the first quarter, Robert's Medical supply paid wages of $360,000. Of this amount, $60,000 was paid to employees who had been paid $7,000 earlier in the quarter. What was the employer's liability for FUTA and SUTA taxes, assuming that the state rate was 5.4%?

SOLUTION: $18,600

$360,000 – $60,000 = $300,000 subject to FUTA and SUTA taxes
$300,000 × 0.008 = $2,400 FUTA tax payment
$300,000 × 0.054 = $16,200 SUTA tax payment
$2,400 + $16,200 SUTA tax payment = $18,600

59.

During the first quarter, Jason Beverage Distributors paid wages of $978,800. Of this amount, $146,000 was paid to employees who had been paid $7,000 earlier in the quarter. What was the employer's liability for FUTA and SUTA taxes, assuming that the state rate was 5.7%? (Round down to the nearest $1.)

SOLUTION: $54,131

$978,800 – $146,000 = $832,800 subject to FUTA and SUTA taxes
$832,800 × 0.008 = $6,662 FUTA tax payment
$832,800 × 0.057 = $47,469 SUTA tax payment
$6,662 + $47,469 SUTA tax payment = $54,131

60.

Kelly Interiors employed Jason Baker for 13 weeks during the first quarter of the year. His salary was $680 per week. Compute the FUTA and SUTA taxes Kelly Interiors must pay on Baker's wages for the quarter, assuming that the state rate was 5.8%?

SOLUTION: $462

$680 per week × 13 weeks = $8,840 total wages
$7,000 maximum × 0.008 = $56 FUTA
$7,000 maximum × 0.058 = $406 SUTA
$56 + $406 = $462 total unemployment taxes

Chapter 11   TAXES

PROBLEMS

LEARNING OBJECTIVE 1

1.
Peter Adams lives in a state having a sales tax of 3%. Compute the amount of tax Peter Adams will pay on a purchase of $590.

SOLUTION: $17.70

2.
Donna Beatty purchased an item costing $160 in a state having a sales tax of 4.5%. Compute the amount of money Donna Beatty paid the sales clerk.

SOLUTION: $167.20

3.
Linda Cloran purchased an item costing $15 in a state having a sales tax of 5%. Compute the amount of change Linda Cloran received from a $20 bill.

SOLUTION: $4.25

4.
David Engler lives in a state having a sales tax of 8%. Compute the amount of tax David Engler will pay on purchases totaling $90.

SOLUTION: $7.20

5.
Paula Jacobs lives in a state having a sales tax rate of 8%. Paula Jacobs crossed the state line into a state having a sales 3% tax rate to make a purchase in the amount of $5,743. Compute the amount Paula Jacobs saved in sales tax by purchasing in the neighboring state.

SOLUTION: $287.15

6.
Steve MacMillan lives in a state having a 6% sales tax. Steve MacMillan purchased items costing the following amounts: $9.20, $6.70, $1.10, $27.50, and $3.50. Compute the total amount of the purchases including tax.

SOLUTION: $50.88

7.
John Naraghi purchased an item costing $18 in a state having a sales tax of 4%. Compute the amount of change John Naraghi received from a $20 bill.

SOLUTION: $1.28

8.
Yvonne Plaatje lives in a state having a sales tax of 6%. Yvonne Plaatje crossed the state line into a state having a 3% tax rate to make a purchase in the amount of $562. If transportation to the next state cost Yvonne Plaatje $20, compute the amount lost by purchasing in another state.

SOLUTION: $3.14

9.
Charles Qualls lives in a state having a sales tax of 6.5%. Compute the amount of tax Charles Qualls will pay on a purchase of $7,400.

SOLUTION: $481

10.
Robert Velasco planned to purchase furniture costing $29,900. The sales tax in Robert Velasco's home state was 8%. The sales tax in a neighboring state was 3%. Delivery charge from the neighboring state would be $1,000. Compute the amount Robert Velasco would save by purchasing furniture from the neighboring state.

SOLUTION: $495

LEARNING OBJECTIVE 2

11.
Indian Bluffs has a valuation rate of 40%. If a building in Indian Bluffs has a market value of $150,000, compute the amount of assessed valuation.

SOLUTION: $60,000

12.
Willow Falls has a valuation rate of 30%. If a building in Willow Falls has an assessed valuation of $90,000, compute the market value of the building.

SOLUTION: $300,000

13.
The town of San Juan has a total assessed valuation of $680,000,000. The amount to be raised by taxation is $12,580,000. Compute the tax rate and express the answer as a percentage.

SOLUTION: 1.85%

14.
Culver City has a tax rate of 17.5 mills per $1. Compute the amount of property tax that will be paid on property having an assessed valuation of $70,000.

SOLUTION: $1,225

15.
The town of Greenville has a total assessed valuation of $540,000,000. The amount to be raised by taxation is $8,640,000. Compute the tax rate and express the answer in mills per dollar.

SOLUTION: 16 mills

16.

Forestville has a total assessed valuation of $340,000,000. The amount to be raised by taxation is $6,290,000. A building in Forestville has an assessed valuation of $120,000. Compute the amount of taxes the building owner will pay.

SOLUTION: $2,220

17.

In a certain development, the houses sell for $250,000. They are located on both sides of a street which is the boundary line between Amber City and Bradley City. Amber City assesses at 40% of market value and has a tax rate of $3.60 per $100. Bradley City assesses at 50% of market value and has a tax rate of $2.98 per $100. Determine which side of the street a buyer would have lower taxes, and by how much.

SOLUTION: Amber City: $125 less

18.

Susan City has a total assessed valuation of $860,000,000. The amount to be raised by taxation is $13,860,000. Compute the tax rate and express the answer as a percentage.

SOLUTION: 1.61%

19.

Hill County is divided into three communities whose assessed valuations, determined individually, are shown in the table below, along with the percentage of assessed value to true value. In the answer column on the right, show what each community's assessment should be for the fair sharing of the county's overhead expenses.

Community	Local Valuation	Rate	Answer
Andersonville	$ 49,120,000	50%	_____
Brownsville	$200,550,000	60%	_____
Carlestown	$138,004,000	75%	_____

SOLUTION:
Andersonville: $ 98,240,000
Brownsville: $334,250,000
Carlestown: $184,005,333

20.

The assessed valuation of a community is 60% and the tax rate is 1.6%. More revenue is needed. The assessed valuation is being left at 60% and the tax rate is being raised to 1.9%. Compute how much more tax money per $100 of market value the increase will generate.

SOLUTION: $0.18

LEARNING OBJECTIVE 3

21.
Express a tax rate of 1.85% in dollars.

SOLUTION: $1.85 on each $100 of value.

22.
Express a tax rate of 4 cents on every dollar in mills.

SOLUTION: 40 mills

23.
Express a tax rate of 19.5 mills as a percent.

SOLUTION: 1.95% 24.

24.
Express a tax rate of 30 mills in pennies on a dollar.

SOLUTION: 3 cents

25.
Fifty mills equal how many pennies?

SOLUTION: 5 cents

26.
Six pennies equal how many mills?

SOLUTION: 60 mills

27.
One dollar equals how many mills?

SOLUTION: 1,000 mills

28.
5,000 mills equal how many dollars?

SOLUTION: $5

29.
Convert 700 mills into cents.

SOLUTION: 70 cents

30.
If a property tax rate is 15 cents on $1, what is that rate in mills?

SOLUTION: 150 mills

LEARNING OBJECTIVE 4

31.
The town of Brownsville assesses property at 80% of market value. The tax rate is 1.8%. A church has a total market value of $580,000. How much does the church save by being exempt from property taxes?

SOLUTION: $8,352

32.
A veteran living in Greenhills receives a partial exemption of 10% of regular property taxes. The veteran owns property valued at $290,000. If the property is assessed at 80% of value and the current rate is 1.4%, how much tax is due each six months?

SOLUTION: $1,461.60

33.
A home having an annual tax bill of $1,800 was sold at the end of the sixth month of the taxable year. The seller had already paid the entire tax for the year. How much tax was the seller reimbursed on proration of taxes at the time of the sale?

SOLUTION: $900

34.
The residents of Ashville voted to widen their roads, at a cost of $220 per residence, with the cost spread over a 10-year period. The Parker family had an annual tax bill of $320 before the improvements. If they pay their property taxes semiannually, what will be the amount of their next tax payment?

SOLUTION: $171

35.
A home having an annual tax bill of $1,500 was sold at the end of the ninth month of the taxable year. The seller had already paid the entire tax for the year. How much tax was the seller reimbursed on proration of taxes at the time of the sale?

SOLUTION: $375

36.
The town of Oakdale assesses property at 70% of market value. The tax rate is 2.3%. A hospital has a total market value of $850,000. How much does the hospital save by being exempt from property taxes?

SOLUTION: $13,685

37.
A veteran living in Tulsa receives a partial exemption of 10% of regular property taxes. The veteran owns property valued at $190,000. If the property is assessed at 70% of value and the current rate is 1.2%, how much tax is due each six months?

SOLUTION: $718.20

38.

A home having an annual tax bill of $2,600 was sold at the end of the ninth month of the taxable year. The seller had already paid the entire tax for the year. How much tax was the seller reimbursed on proration of taxes at the time of the sale?

SOLUTION: $650

39.

The residents of Denton voted to build a new library, at a cost of $120 per residence, with the cost spread over a 10-year period. The McCracken family had an annual tax bill of $1,240 before the improvements. If they pay their property taxes semiannually, what will be the amount of their next tax payment?

SOLUTION: $626

40.

A home having an annual tax bill of $2,400 was sold at the end of the tenth month of the taxable year. The seller had already paid the entire tax for the year. How much tax was the seller reimbursed on proration of taxes at the time of the sale?

SOLUTION: $400

LEARNING OBJECTIVE 5

41.

Margaret Rose works as a waitress. Last year she earned $23,500 in salary, $8,700 in tips, and $1,500 catering on weekends. Compute her gross income.

SOLUTION: $33,700

42.

Joe and Amy Kivers both worked last year. Joe earned $48,000 as a mechanic. Amy earned $18,000 doing typing at home. They inherited $12,000. Compute their combined gross income.

SOLUTION: $66,000

43.

James and Judy Smith own a small apartment complex. They received rents of $26,000 last year. Judy painted and sold landscapes for $11,700. Tom received $9,300 in royalties for some magazine articles. Compute their combined gross income.

SOLUTION: $47,000

44.

Henry Walton received $22,000 as manager of an apartment building, $50,000 as beneficiary of his late wife's insurance policy, $5,400 in interest on tax-exempt state bonds, and a gift of $900. Compute Henry's gross income.

SOLUTION: $22,000

45.

Miriam Jordan is a single individual taking the standard deduction. The normal standard deduction for a single taxpayer is $4,850. She is 72 years old. Compute Miriam's standard deduction.

SOLUTION: $6,050

46.

Jake and Sharon Lewis are married, filing jointly, and taking the standard deduction. Sharon is 64 years old. Jake is 66 years old. The normal standard deduction for a joint return is $9,700. Compute their standard deduction.

SOLUTION: $10,650

47.

Justin Church is a single individual taking the standard deduction. He is 81 years old and blind. The normal standard deduction for a single taxpayer is $4,850. Compute Church's standard deduction.

SOLUTION: 7,250

48.

Harold and Shawna Kaplan are married, filing jointly, and taking the standard deduction. Harold is 67 years old. Shawna is 66 years old. The normal standard deduction for a joint return is $9,700. Compute their standard deduction.

SOLUTION: $11,600

49.

Ellery and Ellen Donaldson are married, filing jointly, and taking the standard deduction. Ellery is 68 years old. Ellen is 65 years old and blind. The normal standard deduction for a joint return is $9,700. Compute their standard deduction.

SOLUTION: $12,550

50.

Elizabeth and Michael Snider had a combined gross income of $58,420. They took the standard deduction of $9,700 for a married couple filing a joint return. They have 3 children ages 5, 7, and 9 years. The amount for each exemption is $3,100. Compute their taxable income.

SOLUTION: $33,220

LEARNING OBJECTIVE 6

51.

Rates for a single taxpayer are 10% of taxable income up to $7,150 and 15% thereafter up to $29,050. Alan James, a single taxpayer, earned $30,000. He took the standard deduction of $4,850 for a single taxpayer and one exemption of $3,100. Compute the amount of income tax.

SOLUTION: $2,950

52.

Rates for a single taxpayer are 10% of taxable income up to $7,150 and 15% thereafter up to $29,050. Ernest Silva, a single taxpayer, earned $25,000 working days as a waiter and $5,000 working nights as a part-time janitor in a local factory. He itemizes the following deductions: $3,800 interest on his house, $1,100 state taxes, $300 donation to his church, and $800 donation to the Salvation Army. He takes one exemption of $3,100. Compute the amount of income tax.

SOLUTION: $2,777.50

53.

Rates for a couple filing a joint tax return are 10% up to $14,300 and 15% thereafter up to $58,100. Joanna and Delbert Linton earned combined gross income of $26,000. They took the standard deduction of $9,700 for a married couple filing a joint return and two exemptions of $3,100 each. Compute the amount of income tax.

SOLUTION: $1,010

54.

Rates for a couple filing a joint tax return are 10% up to $14,300 and 15% thereafter up to $58,100. Samuel and Sharon Duncan earned combined gross income of $50,000. They took the standard deduction of $9,700 for a married couple filing a joint return and two exemptions of $3,100 each. Compute the amount of income tax.

SOLUTION: $4,400

55.

Rates for a couple filing a joint tax return are 10% up to $14,300 and 15% thereafter up to $58,100. William and Dana Souza earned combined gross income of $67,000. They took the standard deduction of $9,700 for a married couple filing a joint return and exemptions of $3,100 each for themselves and their three children. Compute the amount of income tax before credits.

SOLUTION: $5,555

56.

Rates for a couple filing a joint tax return are 10% up to $14,300 and 15% thereafter up to $58,100. Sadie and Gilbert Goldberg earned combined gross income of $72,000. They itemized the following deductions: $5,400 interest on their home, $4,700 state taxes, $2,500 donation to their church, and a $700 donation to the Salvation Army. They took exemptions of $3,100 each for themselves and their two children. Compute the amount of income tax before credits.

SOLUTION: $6,230
$72,000 - (5400+4700+2500+700) = 58700 - 12400 = 46300.   46300 - 14300 = 32000 × 0.15 = 4800 + 1430 = 6230

57.

Rates for a head-of-household taxpayer are 10% of taxable income up to $10,200 and 15% thereafter up to $38,900. Delbert Canfield, head of household, earned $38,000. He took the standard deduction of $7,150 for a head of household and exemptions of $3,100 each for himself and his two children. Compute the amount of income tax before credits.

SOLUTION: $2,722.50

58.

Rates for a head-of-household taxpayer are 10% of taxable income up to $10,200 and 15% thereafter up to $38,900. Earlene Brown, head of household, earned $41,700. She itemized the following deductions: $4,600 interest on her house, $2,100 state tax, and a donation of $575 to her church. She took exemptions of $3,100 each for herself and her daughter. Compute the amount of income tax before credits.

SOLUTION: $3,723.75

59.

Rates for a married taxpayer filing separately are 10% of taxable income up to $7,150 and 15% thereafter up to $29,050. Murry and Ann Phillips are filing separate returns. Murry earned $32,000 this year. He took the standard deduction of $4,850 and exemptions of $3,100 each for himself and the three children. Ann earned $21,000 this year. She took the standard deduction of $4,850 and an exemption of $3,100 for herself. Compute the amount of tax that the Phillips family owed this year before credits.

SOLUTION: $1,855 + 1,600 = $3,455

60.

Rates for a married taxpayer filing separately are 10% of taxable income up to $7,150 and 15% thereafter up to $29,050. Rates for a couple filing a joint return are 10% of taxable income up to $14,300 and 15% thereafter up to $58,100. Patrick and Kelly O'Day are figuring their tax both ways for comparison before deciding which way to file. Patrick earned $20,000 and Kelly earned $35,000. The standard deduction for a married couple filing jointly is $9,700. The standard deduction for each married taxpayer filing separately is $4,850. Exemption for each taxpayer is $3,100.

a.   Compute the amount of tax the O'Days will owe if they file a joint return.
b.   Compute the amount of tax the O'Days will owe if they file two separate returns.

SOLUTION: a. $5,150
                       b. $1,450 + $3,700 = $5,150

Chapter 12 INSURANCE

PROBLEMS

LEARNING OBJECTIVE 1

1.
Thomas Insurance Company does business in a state having no-fault insurance. Driver Jones is insured by Thomas Insurance Company and carries only no-fault insurance. Jones skids on a curve and hits a fruit stand. Medical expenses for Jones are $720 and are $630 for the only passenger. Repairs to the car cost $1,300. Repairs for the fruit stand cost $420. Compute the amount Jones and Thomas Insurance Company each had to pay toward repairs and medical expenses.

SOLUTION: Jones: $1,720; Thomas Insurance Company: $1,350

2.
Drivers Martin and Stein live in a state having no-fault auto insurance. Stein causes an accident by hitting Martin's car. Stein is not hurt. Martin spends three days in the hospital at a cost of $3,300. Compute the amount each driver's insurance company pays toward medical expenses.

SOLUTION: Martin's company: $3,300; Stein's company: $0

3.
Driver Buckley lives in a state having no-fault auto insurance. Goodroads Insurance Company insures Buckley for no-fault insurance only. Buckley's car is struck by a truck while waiting at a stop light. Medical expenses for Buckley and passenger are $547. Repairs to Buckley's car cost $320. Six months later, a car backs out of an alley and hits Buckley's car, causing medical expenses of $950 for Buckley and passenger and $1,249 for repairs to Buckley's car. Compute the amount Goodroads Insurance Company pays for Buckley's involvement in accidents this year.

SOLUTION: $1,497

4.
Safeco Insurance Company insured Driver Fitzwater at an annual premium of $3,300. After one month, Fitzwater sold the car and canceled the insurance. Safe Insurance Company refunded the remaining 11 months' premium at the short rate based on a 15% penalty. Compute the amount that the one month of insurance cost Fitzwater.

SOLUTION: $770

5.
Driver Sanders has a poor driving record and pays triple the regular premium. The regular premium would be $2,100 and covers Sanders for all three classifications of insurance with a $500 deductible for auto collision. During the year, Sanders sideswiped a row of parked cars. Repairs to the parked cars cost $1,870, $4,950, $2,340, and $378. Repairs to Sanders' car cost $490. Compute the amount the insurance company paid out in benefits over the amount it received in premium for this policy.

SOLUTION: $3,238

6.

Manley Insurance Company does business in a state having no-fault insurance. An insured carries all classifications of insurance, with a $500 deductible for collision. The insured paid an annual premium of $2,770. The insured struck a telephone pole causing medical expenses of $2,340 for his passenger, $620 damage to his car, and $1,840 damage to property. How much more did Manley Insurance Company pay out on behalf of the insured than it received as his annual premium?

SOLUTION: $1,530

7.

Universal Insurance Company does business in a state having no-fault insurance. An insured carries all three classifications of insurance, with a $500 deductible for collision. The insured paid an annual premium of $2,790. The insured fell asleep at the wheel and had medical expenses for him self of $1,280 and auto repair expenses of $800. Compute the amount that Universal Insurance Company premium was over the amount the insurance company paid in benefits.

SOLUTION: $1,210

8.

An insured carries all three classifications of insurance, with a deductible of $500 for collision. The insured paid an annual premium of $1,720. The insured backed into a car in a parking lot, causing $580 damage to the car of the insured and $1,770 damage to the other car. Compute the amount that the insurance company made from the insured this year.

SOLUTION: $0

LEARNING OBJECTIVE 2

9.

Driver Smith is given a 15% discount on auto insurance as a low-risk driver. The regular premium would be $2,300 for a year. If Smith qualifies for the discount every year for 5 years, compute the amount Smith will save on insurance premiums.

SOLUTION: $1,725.

10.

Driver Cooper has a poor driving record and pays double the usual premium as a high-risk driver. The regular premium would be $510 for a year. If Cooper must pay the high-risk premium every year for 5 years, how much more will Cooper pay for insurance premiums than a low-risk driver receiving a 10% discount over the same 5-year period?

SOLUTION: $2,805

11.

Driver Blake, a low-risk driver, is insured by a company that gives a 5% discount for the first year, 10% the second year, 15% the third year, and 20% every year thereafter. Blake has qualified as a low-risk driver for the last 10 years. Compute the amount Blake saved in premiums if the regular premium is $2,000.

SOLUTION: $3,400

LEARNING OBJECTIVE 3

12.

A homeowner insured a house for $150,000 for one year at a premium rate of $5.00 per thousand. The homeowner canceled the policy during the year. The insurance company has a penalty of 15% for short-rate refunds. Compute the amount of the penalty.

SOLUTION: $112.50

13.

Cameron Insurance Company insured Driver Gifford at an annual premium of $740. After 6 months, Gifford sold the car and canceled the insurance. Cameron Insurance Company refunded the remaining half of the premium at the short rate based on a penalty of 15%. Compute the amount of the short-rate refund.

SOLUTION: $259

14.

Fuller Insurance Company insured Driver Hudson at an annual premium of $800. After 2 months, Fuller Insurance Company canceled the policy. Compute the amount of the refund to Hudson.

SOLUTION: $666.67

15.

Garnet Corporation insured a building for $370,000 for one year at a premium rate of $6.50 per thousand. Three months later the Garnet Corporation sold the building and canceled the policy. The insurance company refunded the remaining three-quarters of the premium at the short-rate based on a penalty of 10% of the annual premium. Compute the amount the insurance company refunded.

SOLUTION: $1,563.25

16.

Cameron Corporation insured a building for $250,000 for one year at a premium rate of $6.30 per thousand. Four months later the Cameron Corporation sold the building and canceled the policy. The insurance company refunded the remaining two-thirds of the premium at the short-rate based on a penalty of 15% of the annual premium. Compute the amount that the four months of insurance cost the Cameron Corporation.

SOLUTION: $761.25

17.

Allied Industries, Inc. insured an office building for $390,000 for one year at a premium rate of $7.10 per thousand. At the end of nine months, the insurance company canceled the policy. Compute the amount the amount of the refund received by Allied Industries, Inc.

SOLUTION: $692.25

18.
Martin Insurance Company issued insurance policies on buildings A and B in the same area for one year at a premium rate of $5.50 per thousand. Building A was insured for $75,000. Building B was insured for $83,000. Martin Insurance Company had a short-rate refund policy based on a penalty of 10% of the annual premium. At the end of the second month, building A was sold and the policy canceled by the building owner. At the end of the sixth month, Martin Insurance Company canceled the insurance on building B. Compute the amount Martin Insurance Company earned altogether by insuring buildings A and B.

SOLUTION: $338.25

19.
Harrington Insurance Company insures four moving vans for National Moving Company. Annual premiums for the four trucks are: Truck A = $3,480; Truck B = $3,360; Truck C = $2,640; and Truck D = $4,100. Harrington Insurance Company charges a short-rate penalty of 15%. At the end of the second month, National Moving Company sold trucks A and B and canceled their insurance. At the end of the fourth month, National Moving Company sold truck C and canceled its insurance. Compute the amount National Moving Company paid Harrington Insurance Company for insurance during the year.

SOLUTION: $7,542

20.
Davis Insurance Company insured Driver Ashley's two cars at annual premiums of $3,400 for car A and $2,440 for car B. Ashley sold both cars after six months and canceled the insurance. The insurance company refunded the remaining six months' premiums at the short rate based on a 15% penalty. Compute the amount of the total refund.

SOLUTION: $2,044
$3400 + 2440 = 5840.   5840 \times .15 = 876$ penalty.
$5840 \times 6/12 = 2920 - 876 = 2044$

LEARNING OBJECTIVE 4

21.
An insurance company has a standard coinsurance clause of 90%. Compute the amount of insurance coverage the owner of a building valued at $360,000 must carry in order to avoid being the bearer of part of the insurance.

SOLUTION: $324,000

22.
Insurance Company A has a standard 90% coinsurance clause for all fire insurance coverage. Insurance Company B has a standard 75% coinsurance clause for all fire insurance coverage. A building is valued at $195,000. How much more insurance coverage would Insurance Company A require than Insurance Company B for full coinsurance coverage?

SOLUTION: $29,250

23.

Property valued at $140,000 was insured for $90,000. The policy contained an 80% coinsurance clause. A fire caused $56,000 in damages. Compute the amount the insurance company paid for repairs.

SOLUTION: $45,000

24.

Property valued at $172,000 was insured for $100,000. The policy contained a 90% coinsurance clause. A fire caused $77,400 in damages. Compute the amount the insurance company paid for repairs.

SOLUTION: $50,000

25.

Property valued at $260,000 was insured for $210,000. The policy contained an 80% coinsurance clause. A fire caused $187,500 in damages. Compute the amount the property owner must pay if the property is repaired for $187,500.

SOLUTION: $187,500

26.

Property valued at $200,000 was insured for $140,000. The policy contained a 60% coinsurance clause. A fire caused $150,000 in damages. Compute the amount the property owner must pay if the property is repaired for $150,000.

SOLUTION: $10,000

27.

The owner of property valued at $150,000 insured the property for $75,000 for one year at a premium rate of $4.90 per thousand. The policy carried a 75% coinsurance clause. A fire caused $56,250 in damages. Compute the amount the property insurance and repairs for fire damage cost the insured that year.

SOLUTION: $19,117.50

28.

The owner of property valued at $140,000 insured the property for $90,000 for one year at a premium rate of $4.80 per thousand. The policy contained a 90% coinsurance clause. A fire caused $63,000 in damage. How much more did the insurance company pay the property owner for repairs for fire damage than the property owner paid the insurance company in premiums that year?

SOLUTION: $44,568

29.

The Faulkner Company building was valued at $450,000. The building was insured for $400,000. The policy contains an 80% coinsurance clause. A fire caused damages of $260,000. Compute the amount the insurance company paid for damages.

SOLUTION: $260,000

30.

The Morgan Company warehouse was valued at $2,400,000. The building was insured for $960,000. The policy contained an 80% coinsurance clause. A fire caused $660,000 in damages. Compute the amount the Morgan Company recovered from the insurance company for fire damage.

SOLUTION: 330,000

31.

The Jordan Company building was valued at $400,000. The building was insured for $200,000. The policy contained an 80% coinsurance clause. A fire caused $180,000 in damages. Compute the amount of the fire damage the Jordan company had to pay.

SOLUTION: $67,500

32.

The Green Company building was valued at $220,000. The building was insured for $100,000 at a premium rate of $5.60 per thousand. The policy contained an 80% coinsurance clause. A small fire caused damages costing $35,200 to repair. Later in the year, another fire caused damages costing $105,600 to repair. How much more did the insurance company pay for repairs for fire damage for Green Company during the year than it earned in premiums from the Green Company for the year?

SOLUTION: $79,440 more

33.

The Norden Appliance Company has two buildings. Building A is valued at $90,000 and is covered by a $60,000 insurance policy with an 80% coinsurance clause. Building B is also valued at $90,000 and is covered by a $50,000 insurance policy with a 75% coinsurance clause. Both buildings burned to the ground. Compute the amount of insurance money Norden Appliance Company collected altogether for fire damages.

SOLUTION: $110,000

LEARNING OBJECTIVE 5

	Straight Life			20-Payment Life			20-Year Endowment		
Age	Annual	Semi-annual	Quarterly	Annual	Semi-annual	Quarterly	Annual	Semi-annual	Quarterly
25	$17.20	$8.94	$4.73	$31.20	$16.26	$8.26	$52.00	$27.04	$14.30
26	17.85	9.28	4.91	31.81	16.52	8.24	52.60	27.35	14.47
27	18.60	9.67	5.11	32.41	16.83	8.64	53.20	27.66	14.63
28	19.30	10.04	5.31	33.06	17.31	8.85	53.86	28.01	14.81

Table 18-1. Typical Premiums for $1,000 Life Insurance

34.

An insured 28 year old purchased a $100,000, 20-payment life policy with premiums payable annually and lived 20 more years. How much more did the insured pay the insurance company in premiums during his lifetime than he would have paid for a $100,000 straight-life policy with premiums payable annually? Refer to Table 18-1. (1 year = 12 months.)

SOLUTION: $27,520

35.

Insured A, age 27, purchased a $35,000, 20-payment life policy with premiums payable annually. Insured B, also age 27, purchased a $35,000 straight-life policy with premiums payable semiannually. Both A and B lived 40 more years. How much more in premiums did insured B pay the insurance company during his lifetime than insured A paid during hers? Refer to Table 18-1. (1 year = 12 months.)

SOLUTION: $4,389

36.

A client, age 26, is planning to purchase a $75,000, 20-payment life policy and is deciding whether to pay quarterly or semiannually. Compute the amount the client would save during her lifetime by choosing semiannual payments. Refer to Table 18-1. (1 year = 12 months.)

SOLUTION: 120
$16.52 \times 75 \times 40 = \$49,560 - 8.24 \times 75 \times 80 = \$49,440$

37.

An insured 25 year old purchased a $100,000, 20-year endowment policy with premiums payable semiannually. Compute the amount the insured paid the insurance company in premiums during his lifetime. Refer to Table 18-1. (1 year = 12 months.)

SOLUTION: $108,160

38.

An insured 27 year old purchased a $60,000, 20-year endowment policy with premiums payable annually. How much more did the insured pay the insurance company in premiums during his lifetime than he would have paid had he chosen a 20-payment life policy for the same amount and with annual premiums? Refer to Table 18-1. (1 year = 12 months.)

SOLUTION: $24,948

39.

An insured 26 year old purchased a $35,000, 20-year endowment policy with premiums payable quarterly. How much more did the insured pay the insurance company during her lifetime than she would have paid had she chosen annual premium payments? Refer to Table 18-1. (1 year = 12 months.)

SOLUTION: $3,696

LEARNING OBJECTIVE 6

Age	Straight Life			20-Payment Life			20-Year Endowment		
	Annual	Semi-annual	Quarterly	Annual	Semi-annual	Quarterly	Annual	Semi-annual	Quarterly
25	$17.20	$8.94	$4.73	$31.20	$16.26	$8.26	$52.00	$27.04	$14.30
26	17.85	9.28	4.91	31.81	16.52	8.24	52.60	27.35	14.47
27	18.60	9.67	5.11	32.41	16.83	8.64	53.20	27.66	14.63
28	19.30	10.04	5.31	33.06	17.31	8.85	53.86	28.01	14.81

Table 18-1. Typical Premiums for $1,000 Life Insurance

End of Policy Year	Straight Life	20-Payment Life	20-Year Endowment
3	$ 15	$ 43	$ 88
4	22	68	130
5	35	93	173
10	104	228	411
15	818	380	684
20	264	552	1,000

Table 18-2. Cash Surrendered and Loan Values of Policies
Issued at age 25 for $1,000

40.

An insured 25 year old purchased a $20,000 straight-life policy. Three years later she needed the maximum loan available on the policy. Compute the amount the insured could borrow. Refer to Tables 18-1 and 18-2. (1 year = 12 months.)

SOLUTION: $300

41.

An insured 25 year old purchased a $35,000, 20-payment life policy. Five years later he needed the maximum loan available on the policy. Compute the amount the insured could borrow. Refer to Tables 18-1 and 18-2. (1 year = 12 months.)

SOLUTION: $3,255

42.

An insured 25 year old purchased a $50,000, 20-year endowment policy. Ten years later he needed to borrow $30,000. After borrowing the maximum on his insurance, how much more did the insured need to borrow elsewhere? Refer to Tables 18-1 and 18-2. (1 year = 12 months.)

SOLUTION: $9,450

43.

An insured 25 year old purchased a $15,000 straight-life policy and a $5,000, 20-payment life policy. Fifteen years later she needed to borrow $2,500 on one of her policies. Determine which policy had a loan value in excess of $2,500. Refer to Tables 18-1 and 18-2. (1 year = 12 months.)

SOLUTION: straight life

44.

An insured 25 year old purchased a $30,000 straight-life policy with annual premiums. Four years later she needed the maximum loan available on the policy. How much more had the insured paid in premiums than she could borrow on the policy? Refer to Tables 18-1 and 18-2. (1 year = 12 months.)

SOLUTION: $1,404 (17.20 × 30 × 4 2,064) – (22 × 30)

45.

An insured 25 year old purchased a $60,000, 20-payment life policy with semiannual premiums. Five years later he needed the maximum loan available on the policy. How much more had the insured paid in premiums than he could borrow on the policy? Refer to Tables 18-1 and 18-2. (1 year = 12 months.)

SOLUTION: $4,176 (16.26 × 60 × 2 × 5) – (93 × 60)

46.

An insured 25 year old purchased a $50,000, 20-year endowment policy with quarterly premiums. Ten years later he needed the maximum loan available on the policy. How much more had the insured paid in premiums than he could borrow on the policy? Refer to Tables 18-1 and 18-2. (1 year = 12 months.)

SOLUTION: $8,050 (14.30 × 50 × 4 × 10) – (411 × 50)

47.

A client, age 25, is considering purchasing a 20-payment life policy with annual payments. He wants to know how much more he will have paid the insurance company in premiums per $1,000 coverage during the 20-year period than he can borrow on the policy at the end of the 20 years. Compute how much more the client will have to pay per thousand. Refer to Tables 18-1 and 18-2. (1 year = 12 months.)

SOLUTION: $72 per thousand more

48.

A client, age 25, is deciding whether to purchase a $25,000, 20-year endowment policy or a $50,000, 20-payment life policy. He plans to build a house 10 years from now and to borrow the maximum available against any policy he owns at that time. How much more could the client borrow on the 20-payment life policy than he could borrow on the 20-year endowment policy? Refer to Table 18-2. (1 year = 12 months.)

SOLUTION: $1,125

LEARNING OBJECTIVE 7

49 – 54.

The Mighty Machinery Company subscribes to the Healthy Family Medical Organization. Monthly insurance premiums are: employee without dependents, $350; employee with one dependent, $450; employee with multiple dependents, $550.

The Mighty Machinery Company pays a large part of the premiums for participating employees and their dependents. Participating employees have deducted from their paychecks every month $40 from an employee without dependents, $80 from an employee with one dependent, and $100 from an employee with multiple dependents.

The Healthy Family Medical Organization has a deductible of $300 per year for an employee with no dependents, $450 per year for an employee with one dependent, and $600 per year for an employee with multiple dependents. The organization pays 80% of all medical bills in excess of the deductible.

To answer all questions regarding health insurance, refer to the above information on Mighty Machinery Company.

49. A single employee participates in the company health plan. Compute the amount deducted from the employee's wages in one year.

50. An employee with 2 dependents participates in the company health plan. Compute the amount deducted from the employee's wages in one year.

51. An employee with nine dependents participates in the company health plan. Compute the amount deducted from the employee's wages in one year.

52. The Mighty Machinery Company has 20 employees. If the company stopped paying for health benefits for employees' dependents, compute the amount the company would pay this year in health premiums to cover only its employees.

53. Adams and his spouse are covered by Mighty Machinery Company's health plan. Compute the total amount of money paid to the insurance company last year by Adams and the Mighty Machinery Company for Adams' health insurance coverage.

54. Brewster, her spouse, and their 5 children are covered by Mighty Machinery Company's health plan. Last year Brewster's doctor bills totaled $5,180. Compute the amount of Brewster's doctor bills the Healthy Family Medical Organization paid.

SOLUTION:
49. $480
50. $1,200
51. $1,200
52. $74,400
53. $5,400
54. $3,664

55 – 60.
The Mighty Machinery Company subscribes to the Healthy Family Medical Organization. Monthly insurance premiums are: employee without dependents, $350; employee with one dependent, $450; employee with multiple dependents, $550.

The Mighty Machinery Company pays a large part of the premiums for participating employees and their dependents. Participating employees have deducted from their paychecks every month $40 from an employee without dependents, $80 from an employee with one dependent, and $100 from an employee with multiple dependents.

The Healthy Family Medical Organization has a deductible of $300 per year for an employee with no dependents, $450 per year for an employee with one dependent, and $600 per year for an employee with multiple dependents. The organization pays 80% of all medical bills in excess of the deductible.

To answer all questions regarding health insurance, refer to the above information on Mighty Machinery Company.

55.   Carlson, his spouse, and their son are covered by Mighty Machinery Company's health plan. Carlson's doctor bills totaled $2,350 last year. Compute Carlson's total medical costs for the year for premiums, deductible, and his percentage of covered medical bills.

56.   The Mighty Machinery Company has 20 employees participating in the health plan. Four employees are single, five have one dependent, and 11 have multiple dependents. Compute the amount the employees' health coverage cost Mighty Machinery Company last year.

57.   Dankins has been employed by the Mighty Machinery Company for one year. He is single and participates in the company health plan. Dankins just had major surgery costing $16,750. How much more money did the insurance company pay out for Dankins than they received in premiums for Dankins' coverage?

58.   Epstein and her spouse are covered by the company health plan. Last year their doctor bills totaled $320. Compute the amount Epstein paid for medical coverage and care last year for herself and one dependent.

59.   Fenway and Gilbert both worked for Mighty Machinery Company two years ago. Each was single and participated in the company health plan individually. Last year they married each other and participated as employee and one dependent. Two years ago each had doctor bills of $500. This year their combined doctor bills were $1,000. How much less did Fenway and Gilbert pay for medical care and coverage last year as a couple than the year before as two individuals?

60.   Hadley, his spouse, and two children are covered by the company health plan. Last year Hadley's doctor bills totaled $3,840. Compute the amount of the doctor bills that Hadley paid.

SOLUTION:
55. $2,150
56. $96,480
57. $8,960
58. $1,280
59. $120
60. $1,248

Chapter 13 SIMPLE INTEREST

PROBLEMS

LEARNING OBJECTIVE 1

1.

Compute the simple interest. If the time is given in months, let one month be 1/12 of a year. Round answers to the nearest cent.

	Principal, Rate, and Time	Interest
a.	$2,600 at 5% for 3 years	_____
b.	$3,950 at 7% for 2.5 years	_____
c.	$ 800 at 13% for 6 months	_____

SOLUTION:
a. Interest: $390.00
b. Interest: $691.25
c. Interest: $ 52.00

2.

Compute the simple interest. If the time is given in months, let one month be 1/12 of a year. Round answers to the nearest cent.

	Principal, Rate, and Time	Interest
a.	$1,500 at 8% for 2 years	_____
b.	$1,800 at 12% for 9 months	_____
c.	$ 757 at 3% for 4 years	_____

SOLUTION:
a. Interest: $240.00
b. Interest: $162.00
c. Interest: $ 90.84

3.

Compute the simple interest. If the time is given in months, let one month be 1/12 of a year. Round answers to the nearest cent.

	Principal, Rate, and Time	Interest
a.	$2,040 at 9% for 1.5 years	_____
b.	$4,200 at 6% for 4 years	_____
c.	$1,640 at 11% for 18 months	_____

SOLUTION:
a. Interest: $ 275.40
b. Interest: $1,008.00
c. Interest: $ 270.60

LEARNING OBJECTIVE 2

4.

Compute the ordinary interest (360-day year) and the total amount of the loan in the following problems. Round answers to the nearest cent.

	Principal, Rate, and Time	Ordinary Interest	Amount
a.	$3,200 at 6% for 120 days	_____	_____
b.	$1,575 at 5% for 75 days	_____	_____
c.	$4,625 at 7% for 270 days	_____	_____

SOLUTION:

a.	Ordinary Interest: $ 64.00; Amount: $3,264.00
b.	Ordinary Interest: $ 16.41; Amount: $1,591.41
c.	Ordinary Interest: $242.81; Amount: $4,867.81

5.

Compute the ordinary interest (360-day year) and the total amount of the loan in the following problems. Round answers to the nearest cent.

	Principal, Rate, and Time	Ordinary Interest	Amount
a.	$3,975 at 5% for 105 days	_____	_____
b.	$1,750 at 7% for 75 days	_____	_____
c.	$4,080 at 9% for 120 days	_____	_____

SOLUTION:

a.	Ordinary Interest: $ 57.97; Amount: $4,032.97
b.	Ordinary Interest: $ 25.52; Amount: $1,775.52
c.	Ordinary Interest: $122.40; Amount: $4,202.40

6.

Compute the ordinary interest (360-day year) and the total amount of the loan in the following problems. Round answers to the nearest cent.

	Principal, Rate, and Time	Ordinary Interest	Amount
a.	$ 950 at 9% for 60 days	_____	_____
b.	$1,800 at 8% for 90 days	_____	_____
c.	$2,400 at 6% for 30 days	_____	_____

SOLUTION:

a.	Ordinary Interest: $ 14.25; Amount: $ 964.25
b.	Ordinary Interest: $ 36.00; Amount: $1,836.00
c.	Ordinary Interest: $ 12.00; Amount: $2,412.00

7.

Compute the ordinary interest (360-day year) and the total amount of the loan in the following problems. Round answers to the nearest cent.

	Principal, Rate, and Time	Ordinary Interest	Amount
a.	$2,775 at 7% for 240 days	_____	_____
b.	$5,000 at 9% for 225 days	_____	_____
c.	$1,150 at 5% for 2 years	_____	_____

SOLUTION:
a.	Ordinary Interest: $129.50; Amount: $2,904.50
b.	Ordinary Interest: $281.25; Amount: $5,281.25
c.	Ordinary Interest: $115.00; Amount: $1,265.00

8.

Compute the ordinary interest (360-day year) and the total amount of the loan in the following problems. Round answers to the nearest cent.

	Principal, Rate, and Time	Ordinary Interest	Amount
a.	$ 835 at 6% for 135 days	_____	_____
b.	$4,500 at 6% for 330 days	_____	_____
c.	$2,500 at 8% for 240 days	_____	_____

SOLUTION:
a.	Ordinary Interest: $ 18.79; Amount: $ 853.79
b.	Ordinary Interest: $247.50; Amount: $4,747.50
c.	Ordinary Interest: $133.33; Amount: $2,633.33

LEARNING OBJECTIVE 3

9.

Compute the exact interest (365-day year) and the total amount of the loan for the following problems. Round answers to the nearest cent.

	Principal, Rate, and Time	Exact Interest	Amount
a.	$3,500 at 9% for 150 days	_____	_____
b.	$2,275 at 5% for 45 days	_____	_____
c.	$1,850 at 7% for 60 days	_____	_____

SOLUTION:
a.	Exact Interest: $129.45; Amount: $3,629.45
b.	Exact Interest: $ 14.02; Amount: $2,289.02
c.	Exact Interest: $ 21.29; Amount: $1,871.29

10.

Compute the exact interest (365-day year) and the total amount of the loan for the following problems. Round answers to the nearest cent.

	Principal, Rate, and Time	Exact Interest	Amount
a.	$2,480 at 7% for 105 days		
b.	$ 925 at 11% for 120 days		
c.	$2,500 at 5% for 30 days		

SOLUTION:

a.	Exact Interest: $49.94; Amount: $2,529.94
b.	Exact Interest: $33.45; Amount: $ 958.45
c.	Exact Interest: $10.27; Amount: $2,510.27

11.

Compute the exact interest (365-day year) and the total amount of the loan for the following problems. Round answers to the nearest cent.

	Principal, Rate, and Time	Exact Interest	Amount
a.	$1,775 at 7% for 270 days		
b.	$3,150 at 9% for 105 days		
c.	$5,600 at 8% for 45 days		

SOLUTION:

a.	Exact Interest: $91.91; Amount: $1,866.91
b.	Exact Interest: $81.55; Amount: $3,231.55
c.	Exact Interest: $55.23; Amount: $5,655.23

12.

Compute the exact interest (365-day year) and the total amount of the loan for the following problems. Round answers to the nearest cent.

	Principal, Rate, and Time	Exact Interest	Amount
a.	$2,750 at 9% for 120 days		
b.	$1,920 at 8% for 135 days		
c.	$1,100 at 6% for 3 years		

SOLUTION:

a.	Exact Interest: $ 81.37; Amount: $2,831.37
b.	Exact Interest: $ 56.81; Amount: $1,976.81
c.	Exact Interest: $198.00; Amount: $1,298.00

13.

Compute the exact interest (365-day year) and the total amount of the loan for the following problems. Round answers to the nearest cent.

	Principal, Rate, and Time	Exact Interest	Amount
a.	$1,250 at 6% for 135 days	_____	_____
b.	$ 800 at 5% for 270 days	_____	_____
c.	$2,640 at 10% for 180 days	_____	_____

SOLUTION:

a.   Exact Interest: $ 27.74; Amount: $1,277.74

b.   Exact Interest: $ 29.59; Amount: $ 829.59

c.   Exact Interest: $130.19; Amount: $2,770.19

LEARNING OBJECTIVE 4

14.

Compute (a) the ordinary interest, (b) the exact interest, and (c) their difference. Round answers to the nearest cent.

Principal, Rate, and Time	Ordinary Interest	Exact Interest	Difference
$2,625 at 5% for 120 days	a. _____	b. _____	c. _____

SOLUTION:

a. $43.75

b. $43.15

c. $ 0.60

15.

Compute (a) the ordinary interest, (b) the exact interest, and (c) their difference. Round answers to the nearest cent.

Principal, Rate, and Time	Ordinary Interest	Exact Interest	Difference
$1,100 at 12% for 120 days	a. _____	b. _____	c. _____

SOLUTION:

a. $44.00

b. $43.40

c. $0.60

16.

Compute (a) the ordinary interest, (b) the exact interest, and (c) their difference. Round answers to the nearest cent.

Principal, Rate, and Time	Ordinary Interest	Exact Interest	Difference
$1,650 at 8% for 30 days	a. _____	b. _____	c. _____

SOLUTION:
a. $11.00
b. $10.85
c. $ 0.15

17.

Compute (a) the ordinary interest, (b) the exact interest, and (c) their difference. Round answers to the nearest cent.

Principal, Rate, and Time	Ordinary Interest	Exact Interest	Difference
$1,350 at 8% for 200 days	a. _____	b. _____	c. _____

SOLUTION:
a. $60.00
b. $59.18
c. $ 0.82

18.

Compute (a) the ordinary interest, (b) the exact interest, and (c) their difference. Round answers to the nearest cent.

Principal, Rate, and Time	Ordinary Interest	Exact Interest	Difference
$3,650 at 8% for 75 days	a. _____	b. _____	c. _____

SOLUTION:
a. $60.83
b. $60.00
c. $0.83

19.

Compute (a) the ordinary interest, (b) the exact interest and (c) their difference. Round answers to the nearest cent.

Principal, Rate, and Time	Ordinary Interest	Exact Interest	Difference
$ 938 at 11% for 270 days	a. _____	b. _____	c. _____

SOLUTION:
a. $77.39
b. $76.32
c. $1.07

LEARNING OBJECTIVE 5

20.
In each problem, (1) compute the actual exact interest (365-day year) and (2) estimate the interest by rounding the principal to the nearest hundred dollars. For each estimate, assume that a year has 360 days and use the given suggestion to create a shortcut.

	Principal, Rate, and Time	Actual Exact Interest	Estimated Interest
a.	$2,500 at 12% for 62 days	_____	_____ (let T = 60 days)
b.	$ 995 at 9.1% for 120 days	_____	_____ (let I = 9%)
c.	$1,225 at 12% for 29 days	_____	_____ (let T = 30 days)

SOLUTION:
a.   Actual Exact Interest: $50.96; Estimated Interest: $50.00
b.   Actual Exact Interest: $29.77; Estimated Interest: $30.00
c.   Actual Exact Interest: $11.68; Estimated Interest: $12.00

21.
In each problem, (1) compute the actual exact interest (365-day year) and (2) estimate the interest by rounding the principal to the nearest hundred dollars. For each estimate, assume that a year has 360 days and use the given suggestion to create a shortcut.

	Principal, Rate, and Time	Actual Exact Interest	Estimated Interest
a.	$1,350 at 8.88% for 120 days	_____	_____ (let I = 9%)
b.	$2,985 at 5.85% for 31 days	_____	_____ (let I = 6%) (let T = 30 days)
c.	$1,775 at 8.15% for 89 days	_____	_____ (let I = 8%) (let T = 90 days)

SOLUTION:
a.   Actual Exact Interest: $39.41; Estimated Interest: $40.50
b.   Actual Exact Interest: $14.83; Estimated Interest: $15.00
c.   Actual Exact Interest: $35.27; Estimated Interest: $36.00

22.

In each problem, (1) compute the actual exact interest (365-day year) and (2) estimate the interest by rounding the principal to the nearest hundred dollars. For each estimate, assume that a year has 360 days and use the given suggestion to create a shortcut. Round answers to the nearest cent.

	Principal, Rate, and Time	Actual Exact Interest	Estimated Interest
a.	$3,600 at 8.12% for 45 days	_____	_____
			(let I = 8%)
b.	$2,515 at 8% for 91 days	_____	_____
			(let T = 90 days)
c.	$3,280 at 11.8% for 60 days	_____	_____
			(let I = 12%)

SOLUTION:
a.   Actual Exact Interest: $36.04; Estimated Interest: $36.00
b.   Actual Exact Interest: $50.16; Estimated Interest: $50.00
c.   Actual Exact Interest: $63.62; Estimated Interest: $66.00

23.

In each problem, (1) compute the actual exact interest (365-day year) and (2) estimate the interest by rounding the principal to the nearest hundred dollars. For each estimate, assume that a year has 360 days and use the given suggestion to create a shortcut. Round answers to the nearest cent.

	Principal, Rate, and Time	Actual Exact Interest	Estimated Interest
a.	$2,250 at 11.9% for 182 days	_____	_____
			(let I = 12%)
			(let T = 180 days)
b.	$ 725 at 8.8% for 119 days	_____	_____
			(let I = 9%)
			(let T = 120 days)
c.	$2,860 at 8.3% for 271 days	_____	_____
			(let I = 8%)
			(let T = 270 days)

SOLUTION:
a.   Actual Exact Interest: $133.51; Estimated Interest: $135.00
b.   Actual Exact Interest: $ 20.80; Estimated Interest: $ 21.00
c.   Actual Exact Interest: $176.25; Estimated Interest: $174.00

LEARNING OBJECTIVE 2

24.
Jorge Garcia was to lend $3,460 to his cousin at 7% simple interest for 270 days. If the cousin repays everything as agreed upon, how much will Jorge receive? (Use a 360-day year.)

SOLUTION:
$3,460 × 0.07 × 270/360 = $181.65; $3,460 + $181.65 = $3,641.65

25.
Walter Fong borrowed $1,175 at 8% simple interest for 180 days. Compute the total amount that Walter will have to pay on the due date. (Use a 360-day year.)

SOLUTION:
$1,175 × 0.08 × 180/360 = $47; $1,175 + $47 = $1,222

26.
Leslie loaned $2,000 to her daughter for four years. She charged her daughter only 4% simple interest. If the daughter repays all interest and principal on time, how much money will Doreen collect?

SOLUTION:
$2,000 × 0.04 × 4 = $320; $2,000 + $320 = $2,320

27.
An investor borrowed $5,000 to buy stock for a gift boutique. The loan was for 90 days at 10% simple interest. Compute the interest that the investor will pay on the due date. (Use a 360-day year.)

SOLUTION:
$5,000 × 0.10 × 90/360 = $125

28.
Devon Gillies is a college student who is working as a laborer for a landscape company during the summer. Before the first paycheck, the owner loaned Devon $1,000 for 30 days at 5% simple interest. Compute the total amount that Devon will need to repay the owner when he gets paid after 30 days. (Use a 360-day year.)

SOLUTION:
$1,000 × 0.05 × 30/360 = $4.17; $1,000 + $4.17 = $1,004.17

29.
A loan officer approved a $1,500 loan to a small business for 60 days. If the simple interest rate was 12%, compute the interest earned on the loan. (Use a 360-day year.)

SOLUTION:
$1,500 × 0.12 × 60/360 = $30

30.

Andrew Merril, owns a mobile sandwich and pretzel stand. Andrew borrows $5,200 for 135 days at 9% simple interest. What total will Andrew need to repay for both interest and principal? (Use a 360-day year.)

SOLUTION:
$5,200 × 0.09 × 135/360 = $175.50; $5,200 + $175.50 = $5,375.50

31.

Deanna and her sister Mollyanne borrowed $1,600 from their mother. Their mother only charged 3% simple interest and the loan was for two years. Compute the total interest and principal that the sisters need to repay their mother. (Use a 360-day year.)

SOLUTION:
$1,600 × 0.03 × 2 = $96; $1,600 + $96 = $1,696

32.

David Riley owns a five-year-old computer game business. Because of slack business, David needs to borrow $8,000. Compute the interest fee if David gets the loan for 120 days at 12% simple interest. (Use a 360-day year.)

SOLUTION:
$8,000 × 0.12 × 120/360 = $320

LEARNING OBJECTIVE 3

33.

Cortez Sheet Metal borrowed $12,250 to make its quarterly payroll tax payment to the government. The loan was for 45 days at a 8% simple interest rate. Compute the amount of interest that it paid for this loan. (Use a 365-day year.)

SOLUTION:
$12,250 × 0.08 × 45/365 = $120.82

34.

For a few years, Karen Williamson and her sister have operated an office services company that caters to local small businesses. Recently they got an opportunity to buy a used high-speed copy machine for $8,500 which they considered an excellent price. To make the purchase quickly, they agreed to pay $1,000 now and borrow $7,500 from the seller for 60 days at 9.5% simple interest. What is the total amount that they will need to repay the seller at the end of the 60 days? (Use a 365-day year.)

SOLUTION:
$7,500 × 0.095 × 60/365 = $117.12; $7,500 + $117.12 = $7,617.12

35.

Norma Nowak has a beauty shop in the shopping center. Last week Norma borrowed $4,250 to buy additional equipment for her shop. The loan was for 200 days at 9.1% simple interest. How much interest must Norma pay when the 200 days have elapsed? (Use a 365-day year.)

SOLUTION:
$4,250 × 0.091 × 200/365 = $211.92

36.
Edison Automotive Machining Company needs to borrow money to repair a broken milling machine. They can borrow $2,500 for 30 days at 6% simple interest. Compute the total amount that Edison will need to repay. (Use a 365-day year.)

SOLUTION:
$2,500 × 0.06 × 30/365 = $12.33; $2,500 + $12.33 + $2,512.33

37.
Melody Diaper Service is a local company that launders and delivers baby diapers. They need to borrow $6,200 to buy some new washing and drying equipment. Their loan is for 180 days at 8% interest. Compute the interest that Melody will need to pay for this loan. (Use a 365-day year.)

SOLUTION:
$6,200 × 0.08 × 180/365 = $244.60

38.
The main refrigerator at Parkside Grille Restaurant had a serious malfunction. Laura Casper, the owner/chef, found a very nice used replacement refrigerator, but needed $4,500 additional cash immediately. She took out a 75-day loan at 7.5% simple interest. Compute the total, interest and principal, that Laura must pay for this loan. (Use a 365-day year.)

SOLUTION:
$4,500 × 0.075 × 75/365 = $69.35; $4,500 + $69.35 = $4,569.35

39.
Barbara Evanston is a novelty shop owner can save $500 by purchasing merchandise now instead of waiting for two months. However, she will have to borrow $3,200 for 60 days and will have to pay 8.5% simple interest. How much will Barbara have to pay in interest to borrow the money? (Use a 365-day year.)

SOLUTION:
$3,200 × 0.085 × 60/365 = $44.71

40.
Shane Morita opened a pet store near a school, hoping that children would see the pets in the window as they walked home. Almost immediately he needed to borrow money to maintain the pets. He got a 60-day loan for $3,600 at 7.2% simple interest. Compute the amount of Shane's total obligation in 60 days. (Use a 365-day year.)

SOLUTION:
$3,600 × 0.072 × 60/365 = $42.61; $3,600 + $42.61 = $3,642.61

LEARNING OBJECTIVE 6

41.
Wendy Ellison needed to borrow $2,500 to pay tuition at her son's school. She borrowed the money from her credit union at 8% ordinary simple interest (360-day year), and at the end of the loan period, repaid a total of $2,600. Compute the length of the loan period. (to the nearest day)

SOLUTION:
T = I ÷ (PR) = $100 ÷ ($2,500 × 0.08) = 0.5 year; 0.5 × 360 = 180 days

42.

Kyle Winston borrowed $4,000 for 90 days. At the due date, he paid $75 interest plus he repaid all of the principal. What was the ordinary simple interest rate (360-day year) that was charged on the loan? (to nearest 1/10 of a percent)

SOLUTION:
R = I ÷ (PT) = $75 ÷ ($4,000 × 90/360) = 0.075 = 7.5%

43.

Paula Frye loaned $1,260 to a friend at 6% ordinary simple interest (360-day year). At the end of the loan period, Paula received the $1,260 plus $9.45 interest. Compute the length of the loan period. (to the nearest day)

SOLUTION:
T = I ÷ (PR) = $9.45 ÷ ($1,260 × 0.06) = 0.125 year; 0.125 × 360 = 45 days

44.

Rick Anderson is willing to lend modest amounts of money to his employees for short terms at 9% ordinary simple interest (360-day year). In May, one employee borrowed some money for 30 days loan. In June, she repaid the loan plus $6.30 interest. Compute the amount that the employee had borrowed.

SOLUTION:
P = I ÷ (RT) = $6.30 ÷ (0.09 × 30/360) = $840

45.

Dennis Rogers and Betty Chin opened a new art gallery. To help them get started, they borrowed $18,500 for one and one-half years at 8% ordinary simple interest (360-day year). Compute the total amount that they will repay.

SOLUTION:
I = PRT = $18,500 × 0.08 × 1.5 = $2,220; $18,500 + $2,220 = $20,720

46.

Nathan Edwards borrowed $1,600 for 270 days to buy new clothes after an apartment fire. Nathan repaid $1,654. Compute the ordinary simple interest rate (360-day year) that was charged. (to nearest 1/10 of a percent)

SOLUTION:
R = I ÷ (PT) = $54 ÷ ($1,600 × 270/360) = 0.045 = 4.5%

47.

Janice Costa bought some power equipment priced at $1,350. The dealer permitted Eve to pay in 60 days, but at a price of $1,389.60. Compute the ordinary simple interest (360-day year) rate that the dealer charged. (to nearest 1/10 of a percent)

SOLUTION:
R = I ÷ (PT) = $39.60 ÷ ($1,350 × 60/360) = 0.176 = 17.6%

48.

Red Winowski invested money for five years because he was guaranteed that he would earn 12% ordinary simple interest (360-day year) per year. At the end of the five years he received all of his original investment plus an additional $1,500 as the interest that he had earned. How much money had Red invested five years earlier?

SOLUTION:
$P = I \div (RT) = \$1,500 \div (0.12 \times 5) = \$2,500$

49.

A florist borrowed money to buy a large quantity of new imported floral art products to sell in her shop. If it cost her $393.60 to borrow the money for 146 days at 12% exact simple interest (365-day year), what was the amount borrowed?

SOLUTION:
$P = I \div (RT) = \$393.60 \div (0.12 \times 146/365) = \$8,200$

50.

A bank made a $12,000 short-term loan to a new book store and earned $187.40 interest in only 60 days. Compute the exact simple interest rate (365-day year) that the bank charged. (to nearest 1/10 of a percent)

SOLUTION:
$R = I \div (PT) = \$187.40 \div (\$12,000 \times 60/365) = 0.095 = 9.5\%$

51.

Willie White borrowed $2,485 for 105 days at 8% exact simple interest (365-day year). Compute the total amount that Willie must repay at the end of the loan period.

SOLUTION:
$I = PRT = \$2,485 \times 0.08 \times 105/365 = \$57.19; \$2,485 + \$57.19 = \$2,542.19$

52.

Franklin Woo loaned money to his daughter, LeAnn, to help her purchase a car. The loan was for two years at 3.5% exact simple interest (365-day year). When LeAnn repaid the loan after two years, the interest amounted to $385. Compute the amount that Franklin had originally loaned to her.

SOLUTION:
$P = I \div (RT) = \$385 \div (0.035 \times 2) = \$5,500$

53.

Nina Lopez calls a finance company to investigate borrowing $2,700 for 60 days. Nina qualifies for a loan at 9.75% exact simple interest (365-day year). If she accepts the loan, compute the total that Nina will have to repay.

SOLUTION:
$I = PRT = \$2,700 \times (0.0975 \times 60/365) = \$43.27; \$2,700 + \$43.27 = \$2,743.27$

54.

Oliver Douglas opened a repair shop for computer printers. Most of his loans were longer term, but he got a short term loan of $840 for office supplies. This loan was at 8.75% exact simple interest (365-day year). On the due date, Oliver repaid a total of $846.04. What was the length of the loan period? (to the nearest day)

SOLUTION:
T = I ÷ (PR); I = $846.04 - $840 = $6.04
T = $6.04 ÷ ($840 × 0.0875) = 0.0822 × 365 = 30.003 or 30 days

55.

Nelda Brown borrowed $1,420 for 120 days to buy some new furniture for her office. At the due date Nelda repaid a total of $1,448. What was the exact simple interest rate (365-day year) that she paid on the loan? (to nearest 1/10 of a percent)

SOLUTION:
$1,448 – $1,420 = $28 interest
R = I ÷ (PT) = $28 ÷ ($1,420 × 120/365) = 0.05998 or 0.06 = 6%

56.

Alexander Carver loaned $3,000 to Bonnie Humphrey for a vacation trip. Compute the loan period if Alexander charged 10% exact simple interest (365-day year) and Bonnie repaid a total of $3,082.19. (to the nearest day)

SOLUTION:
T = I ÷ (PR) = $82.19 ÷ ($3,000 × 0.10) = 0.274 year; 0.274 × 365 = 100 days

Chapter 14 INSTALLMENT PURCHASES

PROBLEMS

LEARNING OBJECTIVE 1

1.
Change the monthly rates to annual rates.

a.   0.75%   =   _____
b.   0.5%   =   _____
c.   0.8%   =   _____

SOLUTION:
a.   0.75% × 12 = 9%
b.   0.5% × 12 = 6%
c.   0.8% × 12 = 9.6%

2.
Change the monthly rates to annual rates.

a.   1%   =   _____
b.   1.75%   =   _____
c.   1/3%   =   _____

SOLUTION:
a.   1% × 12 = 12%
b.   1.75% × 12 = 21%
c.   1/3% × 12 = 4%

3.
Change the monthly rates to annual rates.

a.   0.6%   =   _____
b.   0.4%   =   _____
c.   1.5%   =   _____

SOLUTION:
a.   0.6% × 12 = 7.2%
b.   0.4% × 12 = 4.8%
c.   1.5% × 12 = 18%

4.
Change the monthly rates to annual rates.

a.   1.2%   =   _____
b.   1.25% =   _____
c.   1.3%   =   _____

SOLUTION:
a.   1.2% × 12 = 14.4%
b.   1.25% × 12 = 15%
c.   1.3% × 12 = 15.6%

5.
Change the annual rates to monthly rates.

a.   6%   =   _____
b.   9%   =   _____
c.   12%   =   _____

SOLUTION:
a.   6% ÷ 12 = 0.5%
b.   9% ÷ 12 = 0.75%
c.   12% ÷ 12 = 1%

6.
Change the annual rates to monthly rates.

a.   21%   =   _____
b.   15%   =   _____
c.   18%   =   _____

SOLUTION:
a.   21% ÷ 12 = 1.75%
b.   15% ÷ 12 = 1.25%
c.   18% ÷ 12 = 1.5%

7.
Change the annual rates to monthly rates.

a.   7.2% =   _____
b.   9.6% =   _____
c.   8.4% =   _____

SOLUTION:
a.   7.2% ÷ 12 = 0.6%
b.   9.6% ÷ 12 = 0.8%
c.   8.4% ÷ 12 = 0.7%

8.

Change the annual rates to monthly rates.

a.   20%   =   _____
b.   16%   =   _____
c.   8%   =   _____

SOLUTION:
a.   20% ÷ 12 = 1 2/3%
b.   16% ÷ 12 = 1 1/3%
c.   8% ÷ 12 = 2/3%

LEARNING OBJECTIVE 2

9.

Convert the annual rate to a monthly rate. Then compute the simple interest on a monthly basis. Round answers to the nearest cent.

Principal, Annual Rate, and Time	Interest
a. $2,600 at 6% for 3 months	_____
b. $ 750 at 7.5% for 7 months	_____
c. $1,800 at 12% for 6 months	_____

SOLUTION:
a.   6% ÷ 12 = 0.5%; $2,600 × 0.005 × 3 = $39.00
b.   7.5% ÷ 12 = 0.625%; $750 × 0.00625 × 7 = $32.81
c.   12% ÷ 12 = 1%; $1,800 × 0.01 × 6 = $108.00

10.

Convert the annual rate to a monthly rate. Then compute the simple interest on a monthly basis. Round answers to the nearest cent.

Principal, Annual Rate, and Time	Interest
a. $5,450 at 9% for 9 months	_____
b. $6,475 at 6.6% for 8 months	_____
c. $ 980 at 7.2% for 5 months	_____

SOLUTION:
a.   9% ÷ 12 = 0.75%; $5, 450 × 0.0075 × 9 = $367.88
b.   6.6% ÷ 12 = 0.55%; $6,475 × 0.0055 × 8 = $284.90
c.   7.2% ÷ 12 = 0.6%; $980 × 0.006 × 5 = $29.40

11.

Convert the annual rate to a monthly rate. Then compute the simple interest on a monthly basis. Round answers to the nearest cent.

	Principal, Annual Rate, and Time	Interest
a.	$2,288 at 15% for 10 months	_____
b.	$3,070 at 10.8% for 4 months	_____
c.	$4,100 at 18% for 2 months	_____

SOLUTION:
a.   15% ÷ 12 = 1.25%; $2,288 × 0.0125 × 10 = $286.00
b.   10.8% ÷ 12 = 0.9%; $2,070 × 0.009 × 4 = $110.52
c.   18% ÷ 12 = 1.5%; $4,100 × 0.015 × 2 = $123.00

12.

Convert the annual rate to a monthly rate. Then compute the simple interest on a monthly basis. Round answers to the nearest cent.

	Principal, Annual Rate, and Time	Interest
a.	$12,500 at 9.6% for 6 months	_____
b.	$10,000 at 8.4% for 3 months	_____
c.	$ 8,775 at 10.5% for 5 months	_____

SOLUTION:
a.   9.6% ÷ 12 = 0.8%; $12,500 × 0.008 × 6 = $600.00
b.   8.4% ÷ 12 = 0.7%; $10,000 × 0.007 × 3 = $210.00
c.   10.5% ÷ 12 = 0.875%; $8,775 × 0.00875 × 5 = $383.91

LEARNING OBJECTIVE 3

13.

Western Farm Machinery has the following credit terms: "The finance charge, if any, is based on the previous balance before payments or credits are deducted. The rates are 1.5% per month up to $1,000 and 1.25% per month on amounts in excess of $1,000. These are annual percentage rates of 18% and 15%, respectively. There is no finance charge if the full amount of the new balance is paid within 30 days after the cycle closing date."

Compute the finance charge and the new balance for the two customers shown below. Assume that both payments were made within the 30-day period.

	Cycle Closing	Previous Balance	Finance Charge	Payments	Purchases	Credits	New Balance
a.	6/20	$2,195.24	_____	$1,400.00	$1,108.42	$150.00	_____
b.	4/20	$ 988.56	_____	$ 500.00	$ 476.15	$ 55.00	_____

SOLUTION:

a.   Previous balance: $2,195.24

                            $- 1,000.00 \times 1.5\% = \$15.00$

                      $\$1,195.24 \times 1.25\% = \underline{+ 14.94}$

                                       $\$29.94$ Finance Charge

      $\$2,195.24 + \$29.94 - \$1,400 - \$150 + \$1,108.42 = \$1,783.60$ New Balance

b.   Previous balance: $\$988.56 \times 1.5\% = \$14.83$ Finance Charge

      $\$988.56 \ \$14.83 - \$500 - \$55 + \$476.15 = \$924.54$ New Balance

14.

Eden Valley Patio Furniture has the following credit terms: "The finance charge, if any, is based on the previous balance before payments or credits are deducted. The rates are 1.5% per month up to $1,000 and 1.25% per month on amounts in excess of $1,000. These are annual percentage rates of 18% and 15%, respectively. There is no finance charge if the full amount of the new balance is paid within 30 days after the cycle closing date."

Compute the finance charge and the new balance for the two customers shown below. Assume that both payments were made within the 30-day period.

	Cycle Closing	Previous Balance	Finance Charge	Payments	Purchases	Credits	New Balance
a.	9/20	$2,009.51	_____	$1,000.00	$788.13	$ 50.00	_____
b.	5/20	$ 837.40	_____	$ 500.00	$372.97	$ 25.00	_____

SOLUTION:

a.   Previous balance: $2,009.51

                          $- 1,000.00 \times 1.5\% \ \ = \$15.00$

                    $\$1,009.51 \times 1.25\% \ = \underline{+ 12.62}$

                                 $\$27.62$ Finance Charge

      $\$2,009.51 + \$27.62 - \$1,000 - \$50 + \$788.13 = \$1,775.26$ New Balance

b.   Previous balance: $\$837.40 \times 1.5\% = \$12.56$ Finance Charge

      $\$837.40 + \$12.56 - \$500 - \$25 + \$372.97 = \$697.93$ New Balance

15.

Turner Landscape Supply, Inc., has the following credit terms: "The finance charge, if any, is based on the previous balance before payments or credits are deducted. The rates are 1.4% per month up to $1,250 and 1.2% per month on amounts in excess of $1,250. These are annual percentage rates of 16.8% and 14.4%, respectively. There is no finance charge if the full amount of the new balance is paid within 30 days after the cycle closing date."

Compute the finance charge and the new balance for the two customers shown below. Assume that both payments were made within the 30-day period.

	Cycle Closing	Previous Balance	Finance Charge	Payments	Purchases	Credits	New Balance
a.	8/25	$3,288.79	_____	$2,500.00	$1,365.83	$347.76	_____
b.	11/25	$1,152.87	_____	$1,000.00	$ 698.44	$ 66.98	_____

SOLUTION:

a.   Previous balance: $3,288.79
                           − 1,250.00 × 1.4%   = $17.50
                           $2,038.79 × 1.2%   = + 24.47
                                                $41.97   Finance Charge

      $3,288.79 + $41.97 − $2,500 + $1,365.83 − $347.76 = $1,848.83 New Balance

b.   Previous balance: $1,152.87 × 1.4% = $16.14 Finance Charge
      $1,152.87 + $16.14 − $1,000 − $66.98 + $698.44 = $800.47 New Balance

16.

Turner Landscape Supply, Inc., has the following credit terms: "The finance charge, if any, is based on the previous balance before payments or credits are deducted. The rates are 1.4% per month up to $1,250 and 1.2% per month on amounts in excess of $1,250. These are annual percentage rates of 16.8% and 14.4%, respectively. There is no finance charge if the full amount of the new balance is paid within 30 days after the cycle closing date."

Compute the payment and purchase history of one customer, Hertzer Landscaping, for two consecutive months. Assume that both payments were made within the 30 day period.

	Cycle Closing	Previous Balance	Finance Charge	Payments	Payment Date	Purchases	Credits	New Balance
a.	6/25	$2,747.51	_____	$2,350	7/20	$842.17	$176.68	_____
b.	7/25	_____	_____	$ 750	8/22	$263.04	$ 19.88	_____

SOLUTION:
a.   Previous balance: $2,747.51
                           − 1,250.00 × 1.4%   = $17.50
                           $1,497.51 × 1.2%   = + 17.97
                                                $35.47   Finance Charge

      $2,747.51 + $35.47 − $2,350 + $842.17 − $176.68 = $1,098.47 New Balance

b. $1,098.47 × 1.4% = $15.38 Finance Charge
      $1,098.47 + $15.38 − $750+ $263.04 − $19.88 = $607.01 New Balance

17.

Bill's Appliance Store offers the following credit terms: "The finance charge is based upon the net balance if payment is made within 25 days of the billing date. If payment is made after 25 days, then the finance charge is based on the previous balance. Net balance equals previous balance less payments, returns and credits. In either case, the monthly interest rate is 1.25% on the first $750 of the net balance and 0.75% on any amount over $750."

Compute the net balance, the finance charge, and the new balance for the two customers shown below. Assume that both payments were made within the 25-day period.

	Previous Balance	Payments	Returns/ Credits	Net Balance	Finance Charge	Purchases	New Balance
a.	$1,808.54	$1,250.00	$411.08	_____	_____	$1,155.90	_____
b.	$2,147.28	$2,027.18	$120.10	_____	_____	$1,614.72	_____

SOLUTION:

a. $1,808.54 – $1,250.00 – $411.08 = $147.46 Net Balance
     $147.46 × 1.25% = $1.84 Finance Charge
     $147.46 + $1.84 + $1,155.90 = $1,305.20 New Balance

b. $2,147.28 – $2,027.18 – $120.10 = $0.00 Net Balance
     $0.00 × 1.25% = $0.00 Finance Charge
     $0.00 + $0.00 + $1,614.72 = $1,614.72 New Balance

18.

Edison Restaurant Supply offers the following credit terms: "The finance charge is based upon the net balance if payment is made within 25 days of the billing date. If payment is made after 25 days, then the finance charge is based on the previous balance. Net balance equals previous balance less payments, returns and credits. In either case, the monthly interest rate is 1.3% on the first $500 of the net balance and 1.1% on any amount over $500."

Compute the net balance, the finance charge, and the new balance for customer Kimberley's Kitchen for two consecutive months. Assume that both payments were made within the 25-day period.

	Previous Balance	Payments	Returns/ Credits	Net Balance	Finance Charge	Purchases	New Balance
a.	$2,874.63	$ 750.00	$261.22	_____	_____	$409.74	_____
b.	_____	$2,000.00	$103.74	_____	_____	$612.77	_____

SOLUTION:

a. $2,874.63 – $750.00 – $261.22 = $1,863.41 Net Balance
     Net Balance: $1,863.41
                  – 500.00 × 1.3% = $ 6.50
                  $1,363.41 × 1.1% = + 15.00
                                         $21.50 Finance Charge
     $1,863.41 + $21.50 + $409.74 = $2,294.65 New Balance

b. $2,294.65 – $2,000.00 – $103.74 = $190.91 Net Balance
     $190.91 × 1.3% = $2.48 Finance Charge
     $190.91 + $2.48 + $612.77 = $806.16 New Balance

LEARNING OBJECTIVES 4, 5

19.

Billy North loaned $2,250 to his former college roommate, Jerold Weinsted. Jerold agreed to repay the principal in three monthly installments of $750 each. Billy charged interest at 0.5% (monthly rate) on the unpaid balance each month. Complete the North-Weinsted loan payment schedule. Then use the North-Weinsted loan payment schedule to solve the equivalent rate problem.

	Month	Unpaid Balance	Interest Payment	Principal Payment	Total Payment	New Balance
a.	1	_____	_____	$750	_____	_____
b.	2	_____	_____	$750	_____	_____
c.	3	_____	_____	$750	_____	_____
	Total	_____	_____			

d. Compute the effective annual interest rate in the North-Weinsted loan agreement by using

$$R = \frac{I}{P \times T}$$

where P is the average principal over the 3-month period, I is the total amount of interest, and T is 3/12 year.

SOLUTION:

a. $2,250 × 0.5% = $11.25 Interest Payment
$750 + $11.25 = $761.25 Total Payment
$2,250 − $750 = $1,500 New Balance

b. $1,500 × 0.5% = $7.50 Interest Payment
$750 + $7.50 = $757.50 Total Payment
$1,500 − $750 = $750 New Balance

c. $750 × 0.5% = $3.75 Interest Payment
$750 + $3.75 = $753.75 Total Payment
$750 − $750 = $0 New Balance

d. $2,250 + $1,500 + $750 = $4,500; P = $4,500 ÷ 3 = $1,500
I = $11.25 + $7.50 + $3.75 = $22.50
R = $22.50 ÷ ($1,500 × 0.25) = $22.50 ÷ $375 = 0.06 or 6%

20.

Anne Harrison borrowed $900 from her uncle to pay off her credit card bill. Anne agreed to repay the principal in three monthly installments of $300 each. Anne's uncle charged interest of 0.75% (monthly rate) on the unpaid balance each month. Complete Anne's loan payment schedule. Then use Anne's loan payment schedule to solve the equivalent rate problem.

	Month	Unpaid Balance	Interest Payment	Principal Payment	Total Payment	New Balance
a.	1	_____	_____	$300	_____	_____
b.	2	_____	_____	$300	_____	_____
c.	3	_____	_____	$300	_____	_____
	Total	_____	_____			

d.   Compute the effective annual interest rate in the Michaelis loan agreement by using

$$R = \frac{I}{P \times T}$$

where P is the average principal over the 3-month period, I is the total amount of interest, and T is 3/12 year.

SOLUTION:

a. $900 × 0.75% = $6.75 Interest Payment
       $300 + $6.75 = $306.75 Total Payment
       $900 − $300   = $600 New Balance

b. $600 × 0.75% = $4.50 Interest Payment
       $300 + $4.50 = $304.50 Total Payment
       $600 − $300   = $300 New Balance

c. $300 × 75%   = $2.25 Interest Payment
       $300 + $2.25 = $302.25 Total Payment
       $300 − $300   = $0 New Balance

d. $900 + $600 + $300 = $1,800; P = $1,800 ÷ 3 = $600
       I = $6.75 + $4.50 + $2.25 = $13.50
       R = $13.50 ÷ ($600 × 3/12) = $13.50 ÷ $150 = 9%

21.

Melinda Paczniak borrowed $4,000 from a private source. Melinda agreed to repay the principal in four installments of $1,000 each. In addition, she paid interest each month, which was calculated by taking 1.5% (monthly rate) of the unpaid balance. Complete Melinda's loan payment schedule. Then use Melinda's loan payment schedule to solve the equivalent rate problem.

	Month	Unpaid Balance	Interest Payment	Principal Payment	Total Payment	New Balance
a.	1	_____	_____	$1,000	_____	_____
b.	2	_____	_____	$1,000	_____	_____
c.	3	_____	_____	$1,000	_____	_____
d.	4	_____	_____	$1,000	_____	_____
	Total	_____	_____			

e.   Compute the effective annual interest rate in the Orebski loan agreement by using

$$R = \frac{I}{P \times T}$$

where P is the average principal over the 4-month period, I is the total amount of interest, and T is 4/12 year.

SOLUTION:

a. $4,000 × 1.5%   = $60 Interest Payment
      $1,000 + $60   = $1,060 Total Payment
      $4,000 − $1,000 = $3,000 New Balance

b. $3,000 × 1.5%   = $45 Interest Payment
      $1,000 + $45   = $1,045 Total Payment
      $3,000 − $1,000 = $2,000 New Balance

c. $2,000 × 1.5%   = $30 Interest Payment
      $1,000 + $30   = $1,030 Total Payment
      $2,000 − $1,000 = $1,000 New Balance

d. $1,000 × 1.5%   = $15 Interest Payment
      $1,000 + $15   = $1,015 Total Payment
      $1,000 − $1,000 = $0 New Balance

e. $4,000 + $3,000 + $2,000 + $1,000 = $10,000
      P = $10,000 ÷ 4 = $2,500
      I = $60 + $45 + $30 + $15 = $150
      R = $150 ÷ ($2,500 × 4/12) = $150 ÷ $833.33 = 18%

22.

As a type of employee fringe benefit, Keith Turner loaned $2,200 to one of his employees. The employee was required to repay the principal in four installments of $550 each. In addition, Turner charged interest each month by taking 0.5% (monthly rate) of the unpaid balance. Complete the loan payment schedule. Then use Turner's loan payment schedule to solve the equivalent rate problem.

	Month	Unpaid Balance	Interest Payment	Principal Payment	Total Payment	New Balance
a.	1	_____	_____	$550	_____	_____
b.	2	_____	_____	$550	_____	_____
c.	3	_____	_____	$550	_____	_____
d.	4	_____	_____	$550	_____	_____
	Total	_____	_____			

e. Compute the effective annual interest rate in the Tagalon loan agreement by using

$$R = \frac{I}{P \times T}$$

where P is the average principal over the 4-month period, I is the total amount of interest, and T is 4/12 year.

SOLUTION:

a.
$2,200 – 0.5% = $11 Interest Payment
$550 + $11 = $561 Total Payment
$2,200 – $550 = $1,650 New Balance

b.
$1,650 × 0.5% = $8.25 Interest Payment
$550 + $8.25 = $558.25 Total Payment
$1,650 – $550 = $1,100 New Balance

c.
$1,100 × 0.5% = $5.50 Interest Payment
$550 + $5.50 = $555.50 Total Payment
$1,100 – $550 = $550 New Balance

d.
$550 × 0.5% = $2.75 Interest Payment
$550 + $2.75 = $552.75 Total Payment
$550 – $550 = $0 New Balance

e.
$2,200 + $1,650 + $1,100 + $550 = $5,500
P = $5,500 ÷ 4 = $1,375
I = $11.00 + $8.25 + $5.50 + $2.75 = $27.50
R = $27.50 ÷ ($1,375 × 4/12) = $27.50 ÷ $458.33 = 6%

LEARNING OBJECTIVES 6, 7

Table 14.1 Amortization Payment Factors per $1,000 Borrowed

Amount of Monthly Payment per $1,000 Borrowed

	Term of Loan	Annual Interest Rate					
		4.5%	6%	7.5%	9%	10.5%	12%
1	month	1003.75000	1005.00000	1006.25000	1007.50000	1008.75000	1010.00000
2	months	502.81425	503.75312	504.69237	505.63200	506.57203	507.51244
3	months	335.83645	336.67221	337.50865	338.34579	339.18361	340.02211
4	months	252.34814	253.13279	253.91842	254.70501	255.49257	256.28109
5	months	202.25561	203.00997	203.76558	204.52242	205.28049	206.03980
6	months	168.86099	169.59546	170.33143	171.06891	171.80789	172.54837
1	Year	85.37852	86.06643	86.75742	87.45148	88.14860	88.84879
2	years	43.64781	44.32061	44.99959	45.68474	46.37604	47.07347
3	years	29.74692	30.42194	31.10622	31.79973	32.50244	33.21431
4	years	22.80349	23.48503	24.17890	24.88504	25.60338	26.33384
5	years	18.64302	19.33280	20.03795	20.75836	21.49390	22.24445
10	years	10.36384	11.10205	11.87018	12.66758	13.49350	14.34709
15	years	7.64993	8.43857	9.27012	10.14267	11.05399	12.00168
20	years	6.32649	7.16431	8.05593	8.99726	9.98380	11.01086
25	years	5.55832	6.44301	7.38991	8.39196	9.44182	10.53224
30	years	5.06685	5.99551	6.99215	8.04623	9.14739	10.28613

23.
Tri-City Financing Corp. amortizes most of the loans that it writes. Use Table 14-1 to compute the monthly payments for the following loans.

a. $2,500 at 10.5% for 4 months _____
b. $8,600 at 12% for 6 months _____
c. $75,000 at 9% for 10 years _____

SOLUTION:

a. $2,500 ÷ $1,000 = 2.5; 2.5 × $255.49257 = $638.73
b. $8,600 ÷ $1,000 = 8.6; 8.6 × $172.54837 = $1,483.92
c. $75,000 ÷ $1,000 = 75; 75 × $12.66758 = $950.07

24.
After getting a bonus, a raise, and a promotion to senior account executive, Daisy Chu bought a new car. With her old car as a trade-in and her bonus as an additional down payment, Daisy needed to borrow $16,000. Her credit union amortized the loan over four years at 9%. Use Table 14-1 to determine the size of Daisy's monthly payments.

SOLUTION:
$16,000 ÷ $1,000 = 16; 16 × $24.88504 = $398.16

25.

Arnold and Lorna Sampson wanted to borrow $35,000 to remodel the kitchen of their home. They were able to get a second mortgage which provided the funds amortized at 9% for 5 years. Use Table 14-1. Compute the size of the Sampson's monthly mortgage payment.

SOLUTION:
$35,000 ÷ $1,000 = 35; 35 × $20.75836 = $726.54

26.

Lawrence and Cynthia Tyler bought a brand new, larger house. They borrowed $275,000 which was to be amortized at 7.5% over 20 years. Use Table 14-1. Compute the size of the Dempsey's monthly mortgage payment.

SOLUTION:
$275,000 ÷ $1,000 = 275; 275 × $8.05593 = $2,215.38

LEARNING OBJECTIVES 5, 6, 7

27.

Eric Russell borrowed $2,400 from a financial loan company which amortized the loan at 12% over 3 months. Using Table 14-1, the first two monthly payments are $816.05. Complete the amortization schedule and solve the effective rate problem.

	Month	Unpaid Balance	Interest Payment	Total Payment	Principal Payment	New Balance
a.	1	_____	_____	$816.05	_____	_____
b.	2	_____	_____	$816.05	_____	_____
c.	3	_____	_____	_____	_____	_____

d. Use Russell's amortization schedule to compute the approximate effective interest rate using

$$R = \frac{I}{P \times T}$$

where P is the average principal over the 3-month period, I is the total amount of interest, and T is 3/12 year.

SOLUTION:
a. Interest Payment: $2,400 × 0.01 = $24.00
    Principal Payment: $816.05 − $24.00 = $792.05
    New Balance: $2,400.00 − $792.05 = $1,607.95

b. Interest Payment: $1,607.95 × 0.01 = $16.08
    Principal Payment: $816.05 − $16.08 = $799.97
    New Balance: $1,607.95 − $799.97 = $807.98

c.   Interest Payment: $807.98 × 0.01 = $8.08
          Total Payment: $807.98 + $8.08 = $816.06
          Principal Payment: $807.98
          New Balance: $807.98 − $807.98 = $0

d.   P = ($2,400 + $1,607.95 + $807.98) ÷ 3 = $1,605.31
          I = $24.00 + $16.08 + $8.08 = $48.16
          R = $48.16 ÷ ($1,605.31 × 3/12) = 0.12000= 12%

28.

Cathy Cortez-Ochoa borrowed $6,000 from her uncle who amortized the loan at 6% over 3 months.

a.   Using Table 14-1, compute the amount of the regular monthly payment. (The last payment may be slightly different.) Complete the amortization schedule and solve the effective rate problem.

	Month	Unpaid Balance	Interest Payment	Total Payment	Principal Payment	New Balance
b.	1	_____	_____	_____	_____	_____
c.	2	_____	_____	_____	_____	_____
d.	3	_____	_____	_____	_____	_____

e.   Use the amortization schedule to compute the effective interest rate using

$$R = \frac{I}{P \times T}$$

where P is the average principal over the three months and I is the total interest charge.

SOLUTION:

a.   Monthly Payment: $6,000 ÷ $1,000 = 6; 6 × 336.67221 = $2,020.03

b.   Interest Payment: $6,000 × 0.005 = $30
          Principal Payment: $2,020.03 − $30 = $1,990.03;
          New Balance: $6,000 − $1,990.03 = $4,009.97

c.   Interest Payment: $4,009.97 × 0.005 = $20.05
          Principal Payment: $2,020.03 − $20.05 = $1,999.98;
          New Balance: $4,009.97 − $1,999.98 = $2,009.99

d.   Interest Payment: $2,009.99 × 0.005 = $10.05
          Total Payment: $2,009.99 + $10.05 = $2,020.04
          Principal Payment: $2,009.99
          New Balance: $2,009.99 − $2,009.99 = $0.00

e.   P = ($6,000 + $4,009.97 + $2,009.99) ÷ 3 = $4,006.65
          I = $30 + $20.05 + $10.05 = $60.10
          R = $60.10 ÷ ($4,006.65 × 3/12) = 0.060000 = 6.00%

29.

Alex Petrovich borrowed $4,800 from a finance company which amortized the loan at 9% over 4 months. Using Table 14-1, the first three monthly payments are $1,222.58. Complete the amortization schedule and solve the effective rate problem.

	Month	Unpaid Balance	Interest Payment	Total Payment	Principal Payment	New Balance
a.	1	_____	_____	$1,222.58	_____	_____
b.	2	_____	_____	$1,222.58	_____	_____
c.	3	_____	_____	$1,222.58	_____	_____
d.	4	_____	_____	_____	_____	_____

e.   Use Petrovich's amortization schedule to compute the effective interest rate using

$$R = \frac{I}{P \times T}$$

where P is the average principal over the four months and I is the total interest charge.

SOLUTION:

a.   Interest Payment: $4,800 × 0.0075 = $36.00
       Principal Payment: $1,222.58 − $36.00 = $1,186.58
       New Balance: $4,800 − $1,186.58 = $3,613.42

b.   Interest Payment: I = $3,613.42 × 0.0075 = 27.10
       Principal Payment: $1,222.58 − $27.10 = $1,195.48
       New Balance: $3,613.42 − $1,195.48 = $2,417.94

c.   Interest Payment: I = $2,417.94 × 0.0075 = $18.13
       Principal Payment: $1,222.58 − $18.13 = $1,204.45;
       New Balance: $2,417.94 − $1,204.45 = $1,213.49

d.   Interest Payment: I = $1,213.49 × 0.0075 = $9.10
       Total Payment: $1,213.49 + $9.10 = $1,222.59
       Principal Payment: $1,213.49
       New Balance: $1,213.49 − $1,213.49 = $0.00

e.   I = $36.00 + $27.10 + $18.13 + $9.10 = $90.33
       P = ($4,800.00 + $3,613.42 + $2,417.94 + $1,213.49) ÷ 4 = $3,011.21
       R = $90.33 ÷ ($3,011.21 × 4/12) = 0.089993 = 9%

30.

Bill and Dick Johnston wanted to borrow $15,500 for four months for their business. A local bank offered to make a loan to be amortized at 12%.

a.   Use Table 14-1 to calculate the amount of the regular monthly payment. (The last payment may be slightly different.) Complete the amortization schedule and solve the effective rate problem.

	Month	Unpaid Balance	Interest Payment	Total Payment	Principal Payment	New Balance
b.	1	_____	_____	_____	_____	_____
c.	2	_____	_____	_____	_____	_____
d.	3	_____	_____	_____	_____	_____
e.	4	_____	_____	_____	_____	_____

f.   Use the Johnston's amortization schedule to compute the approximate effective interest rate using

$$R = \frac{I}{P \times T}$$

where P is the average principal over the four months and I is the total interest charge.

SOLUTION:

a.   Monthly Payment: $15,500 ÷ $1,000 = 15.5; 15.5 × 256.28109 = $3,972.36

b.   Interest Payment: I = $15,500 × 0.01 = $155.00
        Principal Payment: $3,972.36 – $155.00 = $3,817.36
        New Balance: $15,500 – $3,817.36 = $11,682.64

c.   Interest Payment: $11,682.64 × 0.01 = $116.82
        Principal Payment: $3,972.36 – $116.82 = $3,855.54
        New Balance: $11,682.64 – $3,855.54 = $7,827.10

d.   Interest Payment: $7,827.10 × 0.01 = $78.27
        Principal Payment: $3,972.36 – $78.27 = $3,894.09
        New Balance: $7,827.10 – $3,894.09 = $3,933.01

e.   Interest Payment: I = $3,933.01 × 0.01 = $39.33
        Total Payment: $3,933.01 + $39.33 = $3,972.34
        Principal Payment: $3,933.01
        New Balance: $3,933.01 – $3,933.01 = $0.00

f.   P = ($15,500.00 + $11,682.64 + $7,827.10 + $3,933.01) ÷ 4 = $9,735.69
        I = $155.00 + $116.82 + $78.27 + $39.33 = $389.42
        R = $389.42 ÷ ($9,735.69 × 4/12) = 0.119998 = 12.00%

Chapter 15 PROMISSORY NOTES AND DISCOUNTING

PROBLEMS

LEARNING OBJECTIVE 1

1.
Compute the number of days upon which interest would be computed in the following situations. Check for leap years.

	From	To	Number of Days
a.	May 7, 2007	September 11, 2007	_____
b.	January 10, 2006	July 12, 2006	_____
c.	December 12, 2005	April 12, 2006	_____

SOLUTION:
a. 127
b. 183
c. 121

2.
Compute the number of days upon which interest would be computed in the following situations. Check for leap years.

	From	To	Number of Days
a.	July 29, 2005	November 7, 2005	_____
b.	August 15, 2008	October 15, 2008	_____
c.	January 16, 2007	April 26, 2007	_____

SOLUTION:
a. 101
b. 61
c. 100

3.
Compute the number of days upon which interest would be computed in the following situations. Check for leap years.

	From	To	Number of Days
a.	February 20, 2006	November 20, 2006	_____
b.	March 7, 2008	July 7, 2008	_____
c.	October 12, 2007	February 22, 2008	_____

SOLUTION:
a. 273
b. 122
c. 133

4.

Compute the number of days upon which interest would be computed in the following situations. Check for leap years.

	From	To	Number of Days
a.	October 9, 2006	April 19, 2007	
b.	September 21, 2007	December 7, 2007	
c.	June 12, 2006	November 3, 2006	

SOLUTION:
a.   192
b.   77
c.   144

LEARNING OBJECTIVE 2

5.

Determine the maturity date for each of the following notes. Consider leap years where necessary.

	Date of Note	Time	Maturity Date
a.	November 7, 2006	120 days	
b.	October 26, 2007	90 days	
c.	May 30, 2005	60 days	

SOLUTION:
a.   March 7, 2007
b.   January 24, 2008
c.   July 29, 2005

6.

Determine the maturity date for each of the following notes. Consider leap years where necessary.

	Date of Note	Time	Maturity Date
a.	July 15, 2005	180 days	
b.	May 24, 2007	4 months	
c.	May 9, 2008	150 days	

SOLUTION:
a.   January 11, 2006
b.   September 24, 2007
c.   October 6, 2008

7.

Determine the maturity date for each of the following notes. Consider leap years where necessary.

	Date of Note	Time	Maturity Date
a.	February 27, 2006	165 days	_____
b.	September 12, 2005	135 days	_____
c.	February 8, 2007	270 days	_____

SOLUTION:
a. August 11, 2006
b. January 25, 2006
c. November 5, 2007

8.

Determine the maturity date for each of the following notes. Consider leap years where necessary.

	Date of Note	Time	Maturity Date
a.	March 25, 2005	105 days	_____
b.	April 14, 2008	100 days	_____
c.	August 31, 2007	6 months	_____

SOLUTION:
a. July 8, 2005
b. July 23, 2008
c. February 28, 2008 Won't 2008 be a leap year? If so, I would have Feb 29 as the due date.

LEARNING OBJECTIVES 1, 2, 3

9.

Compute the maturity value on each of the following notes. Fill in the missing entries for time or maturity date. (Use a 360-day year.)

	Face Value	Date	Time	Due Date	Rate	Maturity Value
a.	$ 9,500	Nov. 13, 2007	90 days	_____	8%	_____
b.	$ 2,450	Apr. 19, 2007	_____	Sep. 19, 2007	9%	_____

SOLUTION:
a. Due Date: Feb. 11, 2008;   Maturity Value: $9,690.00
b. Time: 153 days;   Maturity Value: $2,543.71

10.

Compute the maturity value on each of the following notes. Fill in the missing entries for time or maturity date. (Use a 360-day year.)

	Face Value	Date	Time	Due Date	Rate	Maturity Value
a.	$12,460	May 12, 2008	_____	Aug. 18, 2008	7%	_____
b.	$ 880	Feb. 3, 2005	120 days	_____	6%	_____

SOLUTION:

a.   Time: 98 days;   Maturity Value: $12,697.43
b.   Due Date: June 3, 2005;   Maturity Value: $ 897.60

11.

Compute the maturity value on each of the following notes. Fill in the missing entries for time or maturity date. (Use a 360-day year.)

	Face Value	Date	Time	Due Date	Rate	Maturity Value
a.	$6,420	Nov. 28, 2005	60 days	_____	5.4%	_____
b.	$2,500	June 7, 2006	_____	Aug. 21, 2006	7.2%	_____

SOLUTION:

a.   Due Date: Jan. 27, 2006; Maturity Value: $6,477.78
b.   Time: 75 days;   Maturity Value: $2,537.50

12.

Compute the maturity value on each of the following notes. Fill in the missing entries for time or maturity date. (Use a 360-day year.)

	Face Value	Date	Time	Due Date	Rate	Maturity Value
a.	$ 8,750	Mar. 21, 2007	_____	Sep. 21, 2007	8.4%	_____
b.	$14,600	Sep. 15, 2006	150 days	_____	9.6%	_____

SOLUTION:

a.   Time: 184 days;   Maturity Value: $ 9,125.67
b.   Due Date: Jan. 28, 2007; Maturity Value: $15,184.00

13.

Compute the maturity value on each of the following notes. Fill in the missing entries for time or maturity date. (Use a 365-day year.)

	Face Value	Date	Time	Due Date	Rate	Maturity Value
a.	$2,470	Oct. 10, 2005	120 days	_____	7%	_____
b.	$5,225	Jan. 14, 2005	_____	Apr. 14, 2005	9%	_____

SOLUTION:

a.   Due Date: Feb. 7, 2006;   Maturity Value: $2,526.84
b.   Time: 90 days;   Maturity Value: $5,340.95

14.

Compute the maturity value on each of the following notes. Fill in the missing entries for time or maturity date. (Use a 365-day year.)

	Face Value	Date	Time	Due Date	Rate	Maturity Value
a.	$ 975	June 20, 2006	_____	Nov. 20, 2006	6%	_____
b.	$1,800	Mar. 22, 2007	75 days	_____	8%	_____

SOLUTION:

a.   Time: 153 days;   Maturity Value: $ 999.52
b.   Due Date: June 5, 2007;   Maturity Value: $1,829.59

15.

Compute the maturity value on each of the following notes. Fill in the missing entries for time or maturity date. (Use a 365-day year.)

	Face Value	Date	Time	Due Date	Rate	Maturity Value
a.	$ 3,860	Nov. 15, 2007	105 days	_____	7.8%	_____
b.	$12,180	July 7, 2008	_____	Nov. 10, 2008	6.3%	_____

SOLUTION:

a.   Due Date: Feb. 28, 2008; Maturity Value: $ 3,946.61
b.   Time: 126 days;   Maturity Value: $12,444.89

16.

Compute the maturity value on each of the following notes. Fill in the missing entries for time or maturity date. (Use a 365-day year.)

	Face Value	Date	Time	Due Date	Rate	Maturity Value
a.	$6,512	Apr. 10, 2007	_____	July 10, 2007	8.25%	_____
b.	$5,000	Oct. 18, 2006	180 days	_____	5.75%	_____

SOLUTION:

a.   Time: 91 days;   Maturity Value: $6,645.94
b.   Due Date: Apr. 16, 2007; Maturity Value: $5,141.78

LEARNING OBJECTIVE 4

17.

Equity Financial Corp. discounted the following two interest-bearing notes. Compute the missing information. Use a 360-day year.

a. | Face Value: | $2,750 | | b. | Face Value: | $3,000 |
|---|---|---|---|---|---|
| Date of Note: | April 16 | | | Date of Note: | October 28 |
| Interest Rate: | 8% | | | Interest Rate: | 9% |
| Time to Run: | 105 days | | | Time to Run: | 45 days |
| Discount Date: | May 31 | | | Discount Date: | November 12 |
| Discount Rate: | 11% | | | Discount Rate: | 12% |
| | | | | | |
| Interest Amount: | _____ | | | Interest Amount: | _____ |
| Maturity Value: | _____ | | | Maturity Value: | _____ |
| Maturity Date: | _____ | | | Maturity Date: | _____ |
| Days of Discount: | _____ | | | Days of Discount: | _____ |
| Discount Amount: | _____ | | | Discount Amount: | _____ |
| Proceeds: | _____ | | | Proceeds: | _____ |

SOLUTION:

a. | Interest Amount: | $64.17 | | b. | Interest Amount: | $33.75 |
|---|---|---|---|---|---|
| Maturity Value: | $2,814.17 | | | Maturity Value: | $3,033.75 |
| Maturity Date: | July 30 | | | Maturity Date: | Dec. 12 |
| Days of Discount: | 60 days | | | Days of Discount: | 30 days |
| Discount Amount: | $51.59 | | | Discount Amount: | $30.34 |
| Proceeds: | $2,762.58 | | | Proceeds: | $3,003.41 |

18.

Guarantee Lending Co. discounted the following two interest-bearing notes. Compute the missing information. Use a 360-day year.

a. | Face Value: | $1,800 | | b. | Face Value: | $925 |
|---|---|---|---|---|---|
| Date of Note: | July 20 | | | Date of Note: | August 18 |
| Interest Rate: | 15% | | | Interest Rate: | 12% |
| Time to Run: | 90 days | | | Time to Run: | 60 days |
| Discount Date: | October 3 | | | Discount Date: | September 27 |
| Discount Rate: | 18% | | | Discount Rate: | 16% |
| | | | | | |
| Interest Amount: | _____ | | | Interest Amount: | _____ |
| Maturity Value: | _____ | | | Maturity Value: | _____ |
| Maturity Date: | _____ | | | Maturity Date: | _____ |
| Days of Discount: | _____ | | | Days of Discount: | _____ |
| Discount Amount: | _____ | | | Discount Amount: | _____ |
| Proceeds: | _____ | | | Proceeds: | _____ |

SOLUTION:

a. | Interest Amount: | $67.50 | | b. | Interest Amount: | $18.50 |
|---|---|---|---|---|---|
| Maturity Value: | $1,867.50 | | | Maturity Value: | $943.50 |
| Maturity Date: | Oct. 18 | | | Maturity Date: | Oct. 17 |
| Days of Discount: | 15 days | | | Days of Discount: | 20 days |
| Discount Amount: | $14.01 | | | Discount Amount: | $8.39 |
| Proceeds: | $1,853.49 | | | Proceeds: | $935.11 |

19.

Carrie Umholtz held a $3,200 105-day note dated July 7, bearing interest at 9.6%. On August 11, Carrie took the note to a finance company which discounted it at 11.3%. Use a 365-day year to find the missing information on the loan.

Interest Amount: _____
Maturity Value: _____
Maturity Date: _____
Days of Discount: _____
Discount Amount: _____
Proceeds: _____

SOLUTION:

Interest Amount:	$88.37
Maturity Value:	$3,288.37
Maturity Date:	Oct. 20
Days of Discount:	70 days
Discount Amount:	$71.26
Proceeds:	$3,217.11

20.

Lillian Barnes made extra money by lending small amounts of money to small businesses. One such loan dated September 9, was for $1,000, for 75 days at 9.2% interest. On October 25, Lillian decided to sell the loan to a lending institution which discounted it 13.4%. Use a 365-day year to find the missing information on the loan.

Interest Amount: _____
Maturity Value: _____
Maturity Date: _____
Days of Discount: _____
Discount Amount: _____

SOLUTION:

Interest Amount:	$18.90
Maturity Value:	$1,018.90
Maturity Date:	Nov. 23
Days of Discount:	29 days
Discount Amount:	$10.85
Proceeds:	$1,008.05

21.

Reed Lyons held a $1,790, 60-day note dated December 22, bearing interest at 9%. On January 11, he took the note to Surety Financing Company which discounted it at 13%. Use a 365-day year to find the missing information on the loan.

Interest Amount:	_____
Maturity Value:	_____
Maturity Date:	_____
Days of Discount:	_____
Discount Amount:	_____
Proceeds:	_____

SOLUTION:

Interest Amount:	$26.48
Maturity Value:	$1,816.48
Maturity Date:	Feb. 20
Days of Discount:	40 days
Discount Amount:	$25.88
Proceeds:	$1,790.60

22.

Deidra Holmes held a $2,800, 90-day note dated April 28, bearing interest at 11%. On June 12, she took the note to Great Northern National Bank which discounted it at 16.5%. Use a 365-day year to find the missing information on the loan.

Interest Amount:	_____
Maturity Value:	_____
Maturity Date:	_____
Days of Discount:	_____
Discount Amount:	_____
Proceeds:	_____

SOLUTION:

Interest Amount:	$75.95
Maturity Value:	$2,875.95
Maturity Date:	July 27
Days of Discount:	45 days
Discount Amount:	$58.50
Proceeds:	$2,817.45

23.

Pierce & Erickson Home Loan discounted the following two non-interest-bearing notes. Compute the missing information. Use a 360-day year.

a.   Face Value: $800   b.   Face Value: $2,400

          Date of Note: October 9   Date of Note: June 5

          Time to Run: 90 days   Time to Run: 75 days

          Discount Date: November 8   Discount Date: July 5

          Discount Rate: 12%   Discount Rate: 9%

          Interest Amount: $0   Interest Amount: $0

          Maturity Value: Maturity Value:

          Maturity Date: Maturity Date:

          Days of Discount: Days of Discount:

          Discount Amount: Discount Amount:

          Proceeds: Proceeds:

SOLUTION:

Interest Amount:	$0		b.	Interest Amount:	$0
Maturity Value:	$800			Maturity Value:	$2,400
Maturity Date:	Jan. 7			Maturity Date:	Aug. 19
Days of Discount:	60 days			Days of Discount:	45 days
Discount Amount:	$16			Discount Amount:	$27
Proceeds:	$784			Proceeds:	$2,373

24.

On October 1, Spencer Worthington loaned $2,000 to an old friend from the army. The note securing the loan was non-interest-bearing and for only 30 days. On October 10, Spencer sold the note to a local finance company which discounted it at 12%. Use a 365-day year to find the missing information on the loan.

         Interest Amount: $0

         Maturity Value:

         Maturity Date:

         Days of Discount:

         Discount Amount:

         Proceeds:

SOLUTION:

Interest Amount:	$0
Maturity Value:	$2,000
Maturity Date:	Oct. 31
Days of Discount:	21 days
Discount Amount:	$13.81
Proceeds:	$1,986.19

25.

Marcia Driscoll held a $2,650, 90-day, non-interest-bearing note dated September 15. On October 15, she took the note to a bank which discounted the note at 13%. Use a 365-day year to find the missing information on the loan.

Interest Amount:	$0
Maturity Value:	
Maturity Date:	
Days of Discount:	
Discount Amount:	
Proceeds:	

SOLUTION:

Interest Amount:	$0
Maturity Value:	$2,650
Maturity Date:	Dec. 14
Days of Discount:	60 days
Discount Amount:	$56.63
Proceeds:	$2,593.37

26.

Annamarie Weymeyer held an $1,500, 120-day, non-interest-bearing note dated July 24. On August 16, she took the note to First Bank of the Plains which discounted the note at 11%. Use a 365-day year to find the missing information on the loan.

Interest Amount:	$0
Maturity Value:	
Maturity Date:	
Days of Discount:	
Discount Amount:	
Proceeds:	

SOLUTION:

Interest Amount:	$0
Maturity Value:	$1,500
Maturity Date:	Nov. 21
Days of Discount:	97 days
Discount Amount:	$43.85
Proceeds:	$1,456.15

27.

Juan Munoz held a $1,170, 45-day, non-interest-bearing note dated March 30. On April 30, he took the note to Southwest Equity Loans which discounted the note at 12%. Use a 365-day year to find the missing information on the loan.

Interest Amount:	$0
Maturity Value:	
Maturity Date:	
Days of Discount:	
Discount Amount:	
Proceeds:	

SOLUTION:

Interest Amount:	$0
Maturity Value:	$1,170
Maturity Date:	May 14
Days of Discount:	14 days
Discount Amount:	$5.39
Proceeds:	$1,164.61

LEARNING OBJECTIVE 5

28.

Thrifty Home Loan Corp. uses the discount method of calculating finance charges. For the two loans below, calculate the discount amount that Thrify is charging and the proceeds that go to the borrower. Then calculate the actual interest rate which is based on proceeds rather than the face value. Use a 360-day year and compute actual rates to the nearest 1/100 of a percent.

	Face Value	Discount Rate	Time	Discount Amount	Proceeds	Actual Interest Rate
a.	$5,000	9%	135 days			
b.	$3,480	8%	270 days			

SOLUTION:

	Discount Amount	Proceeds	Actual Interest Rate
a.	$168.75	$4,831.25	9.31%
b.	$208.80	$3,271.20	8.51%

29.

On some personal loans, Buckhorn County Bank & Trust uses the discount method of calculating finance charges. For the two loans below, calculate the discount amount that Buckhorn Bank is charging and the proceeds that go to the borrower. Then calculate the actual interest rate which is based on proceeds rather than the face value. Use a 360-day year and compute actual rates to the nearest 1/100 of a percent.

	Face Value	Discount Rate	Time	Discount Amount	Proceeds	Actual Interest Rate
a.	$6,500	9%	90 days	_____	_____	_____
b.	$1,925	7	150 days	_____	_____	_____

SOLUTION:

	Discount Amount	Proceeds	Actual Interest Rate
a.	$146.25	$6,353.75	9.21%
b.	$ 56.15	$1,868.85	7.21%

30.

Silver Falls Finances Inc. made the following two loans using a discount rate to compute the charge and then loaned the proceeds to their clients. Compute the amount of discount, the proceeds, and the actual interest rate based upon the proceeds rather than the face value. Use a 360-day year and compute actual rates to the nearest 1/100 of a percent.

	Face Value	Discount Rate	Time	Discount Amount	Proceeds	Actual Interest Rate
a.	$2,180	10%	75 days	_____	_____	_____
b.	$3,054	9%	60 days	_____	_____	_____

SOLUTION:

	Discount Amount	Proceeds	Actual Interest Rate
a.	$45.42	$2,134.58	10.21
b.	$45.81	$3,008.19	9.14%

31.

White River Bank made the following two loans using a discount rate to compute the charge and then loaned the proceeds to their clients. Compute the amount of discount, the proceeds, and the actual interest rate based upon the proceeds rather than the face value. Use a 360-day year and compute actual rates to the nearest 1/100 of a percent.

	Face Value	Discount Rate	Time	Discount Amount	Proceeds	Actual Interest Rate
a.	$4,090	11%	120 days	_____	_____	_____
b.	$3,260	10%	150 days	_____	_____	_____

SOLUTION:

	Discount Amount	Proceeds	Actual Interest Rate
a.	$149.97	$3,940.03	11.42%
b.	$135.83	$3,124.17	10.43%

32.

Sincere Personal Loan Corp. uses the discount method of calculating finance charges on many different loans. For the two loans below, calculate the discount amount that goes to Sincere and the proceeds that go to the borrower. Then calculate the actual interest rate which is based on proceeds rather than the face value. Use a 365-day year and compute actual rates to the nearest 1/100 of a percent.

	Face Value	Discount Rate	Time	Discount Amount	Proceeds	Actual Interest Rate
a.	$1,288	10.4%	100 days	_____	_____	_____
b.	$5,840	8.3%	225 days	_____	_____	_____

SOLUTION:

	Discount Amount	Proceeds	Actual Interest Rate
a.	$ 36.70	1,215.3	10.7 1%
b.	$298.80	$5,541.20	8.75%

33.

Southeast Home Loan Bank uses the discount method of calculating finance charges on some loans. For the two loans below, calculate the discount amount that goes to Southeast and the proceeds that go to the borrower. Then calculate the actual interest rate which is based on proceeds rather than the face value. Use a 365-day year and compute actual rates to the nearest 1/100 of a percent.

	Face Value	Discount Rate	Time	Discount Amount	Proceeds	Actual Interest Rate
a.	$3,250	7.8%	90 days	_____	_____	_____
b.	$2,195	9.1%	135 days	_____	_____	_____

SOLUTION:

	Discount Amount	Proceeds	Actual Interest Rate
a.	$62.51	$3,187.49	7.95%
b.	$73.88	$2,121.12	9.42%

34.

Moneywise Loans and Credit Corp. makes many consumer loans by discounting the face value of the loan. The borrower gets the proceeds. For the two loans below, determine the amount of the discount, the proceeds, and the actual interest rate which is based on the proceeds rather than the face value. Use a 365-day year and compute actual rates to the nearest 1/100 of a percent.

	Face Value	Discount Rate	Time	Discount Amount	Proceeds	Actual Interest Rate
a.	$14,800	6.9%	92 days	_____	_____	_____
b.	$10,000	10.3%	180 days	_____	_____	_____

SOLUTION:

	Discount Amount	Proceeds	Actual Interest Rate
a.	$257.40	$14,542.60	7.02%
b.	$507.95	$ 9,492.05	10.85%

35.

First Republic Bank makes many loans to individuals by discounting the face value of the loan. The borrower gets the proceeds. For the two loans below, determine the amount of the discount, the proceeds, and the actual interest rate which is based on the proceeds rather than the face value. Use a 365-day year and compute actual rates to the nearest 1/100 of a percent.

	Face Value	Discount Rate	Time	Discount Amount	Proceeds	Actual Interest Rate
a.	$6,212	8.5%	240 days	_____	_____	_____
b.	$4,885	7.7%	120 days	_____	_____	_____

SOLUTION:

	Discount Amount	Proceeds	Actual Interest Rate
a.	$347.19	$5,864.81	9.00%
b.	$123.66	$4,761.34	7.90%

LEARNING OBJECTIVE 6

36.

Bovio Electrical Contracting tries to take advantage of cash discounts, even if it must borrow the money for a few days in order to pay the invoice in time. Compute the savings on Bovio's purchases and terms from two different suppliers. Use a 360-day year. Assume that the number of interest days is the time between the due date and the last date to take advantage of the cash discount.

	Invoice	Terms	Cash Discount	Interest Rate on Loan	Days of Interest	Amount of Interest	Savings
a.	$3,480	2.5/5, n/45	_____	8%	_____	_____	_____
b.	$ 960	3/5, n/25	_____	9%	_____	_____	_____

SOLUTION:

	Cash Discount	Days of Interest	Amount of Interest	Savings
a.	$87.00	40	$30.16	$56.84
b.	$28.80	20	$ 4.66	$24.14

37.

Garcia Heating & Air Conditioning tries to take advantage of cash discounts, even if it must borrow the money for a few days in order to pay the invoice in time. Compute the savings on Garcia's purchases and terms from two different suppliers. Use a 360-day year. Assume that the number of interest days is the time between the due date and the last date to take advantage of the cash discount.

	Invoice	Terms	Cash Discount	Interest Rate on Loan	Days of Interest	Amount of Interest	Savings
a.	$1,200	1/10, n/25	_____	8%	_____	_____	_____
b.	$4,895	2/20, n/40	_____	11%	_____	_____	_____

SOLUTION:

	Cash Discount	Days of Interest	Amount of Interest	Savings
a.	$12.00	15	$ 3.96	$ 8.04
b.	$97.90	20	$29.32	$68.58

38.

Wei Fang Food Importers often tries to take advantage of cash discounts, even if it must borrow the money for a few days in order to pay the invoice in time. Compute the savings on Wei Fang's purchases and terms from two different suppliers. Use a 365-day year. Assume that the number of interest days is the time between the due date and the last date to take advantage of the cash discount.

	Invoice	Terms	Cash Discount	Interest Rate on Loan	Days of Interest	Amount of Interest	Savings
a.	$2,650	3/5, n/15	_____	9%	_____	_____	_____
b.	$3,600	2/10, n/30	_____	10%	_____	_____	_____

SOLUTION:

	Cash Discount	Days of Interest	Amount of Interest	Savings
a.	$79.50	10	$ 6.34	$73.16
b.	$72.00	20	$19.33	$52.66

39.

Gupta & Grant Computer Supplies usually tries to take advantage of cash discounts, even if it must borrow the money for a few days in order to pay the invoice in time. Compute the savings on Gupta & Grant's purchases and terms from two different suppliers. Use a 365-day year. Assume that the number of interest days is the time between the due date and the last date to take advantage of the cash discount.

	Invoice	Terms	Cash Discount	Interest Rate on Loan	Days of Interest	Amount of Interest	Savings
a.	$1,725	2/15, n/60	_____	9.5%	_____	_____	_____
b.	$2,200	1/10, n/30	_____	8.9%	_____	_____	_____

SOLUTION:

	Cash Discount	Days of Interest	Amount of Interest	Savings
a.	$34.50	45	$19.80	$14.70
b.	$22.00	20	$10.62	$11.38

Chapter 16 COMPOUND INTEREST

Note 1: Tables 16-1A, 16-1B, 16-2A, and 16-2B appear at the end of this chapter.
Note 2: Solutions to present value problems are shown using only Tables 16-2A and 16-2B.
Note 3: If a calculator is used to find the future or present value factor, the answer may vary slightly.

PROBLEMS

LEARNING OBJECTIVE 1

1.
Compute the future value and the compound interest earned for each of the following investments. Use Tables 16-1A&B or a calculator.

	Principal	Interest Rate	Time	Future Value	Compound Interest
a.	$5,500	8% compounded quarterly	3 years	_____	_____
b.	$1,000	10% compounded annually	15 years	_____	_____

SOLUTION:
a. $0.08 \div 4 = 0.02$; $4 \times 3 = 12$; $5,500 \times 1.26824 = \$6,975.32$ Future Value
    $6,975.32 - \$5,500 = \$1,475.32$ Compound Interest
b. $0.10 \div 1 = 0.10$; $1 \times 15 = 15$; $1,000 \times 4.17725 = \$4,177.25$ Future Value
    $4,177.25 - \$1,000 = \$3,177.25$ Compound Interest

2.
Compute the future value and the compound interest earned for each of the following investments. Use Tables 16-1A&B or a calculator.

	Principal	Interest Rate	Time	Future Value	Compound Interest
a.	$8,200	18% compounded monthly	6 months	_____	_____
b.	$3,760	10% compounded semiannually	10 years	_____	_____

SOLUTION:
a. $0.18 \div 12 = 0.015$; 6 months; $8,200 \times 1.09344 = \$8,966.21$ Future Value
    $8,966.2 - \$8,200 = \$766.21$ Compound Interest

b. $0.10 \div 2 = 0.05$; $2 \times 10 = 20$; $3,760 \times 2.65330 = \$9,976.41$ Future Value
    $9,976.41 - \$3,760 = \$6,216.41$ Compound Interest

3.

Compute the future value and the compound interest earned for each of the following investments. Use Tables 16-1A&B or a calculator.

	Principal	Interest Rate	Time	Future Value	Compound Interest
a.	$25,000	9% compounded annually	5 years	_____	_____
b.	$ 2,450	8% compounded semiannually	7 years	_____	_____

SOLUTION:

a. $0.09 \div 1 = 0.09$; $1 \times 5 = 5$; $25,000 \times 1.53862 = $38,465.50 Future Value
        $38,465.50 - $25,000 = $13,465.50 Compound Interest

b. $0.08 \div 2 = 0.04$; $2 \times 7 = 14$; $2,450 \times 1.73168 = $4,242.62 Future Value
        $4,242.62 - $2,450 = $1,792.62 Compound Interest

4.

Compute the future value and the compound interest earned for each of the following investments. Use Tables 16-1A&B or a calculator.

	Principal	Interest Rate	Time	Future Value	Compound Interest
a.	$12,500	12% compounded quarterly	3 years	_____	_____
b.	$ 4,800	9% compounded monthly	2/3 year	_____	_____

SOLUTION:

a. $0.12 \div 4 = 0.03$; $4 \times 3 = 12$; $12,500 \times 1.42576 = $17,822 Future Value
        $17,822 - $12,500 = $5,322 Compound Interest

b. $0.09 \div 12 = 0.0075$; $12 \times 2/3 = 8$; $4,800 \times 1.06160 = $5,095.68 Future Value
        $5,095.68 - $4,800 = $295.68 Compound Interest

5.

Compute the future value and the compound interest earned for each of the following investments. Use Tables 16-1A&B or a calculator.

	Principal	Interest Rate	Time	Future Value	Compound Interest
a.	$1,350	9% compounded annually	22 years	_____	_____
b.	$2,400	8% compounded semiannually	13 years	_____	_____

SOLUTION:

a. $0.09 \div 1 = 0.09$; $1 \times 22 = 22$; $1,350 \times 6.65860 = $8,989.11 Future Value
        $8,989.11 - $1,350 = $7,639.11 Compound Interest

b. $0.08 \div 2 = 0.04$; $2 \times 13 = 26$; $2,400 \times 2.77247 = $6,653.93 Future Value
        $6,653.93 - $2,400 = $4,253.93 Compound Interest

6.

Compute the future value and the compound interest earned for each of the following investments. Use Tables 16-1A&B or a calculator.

	Principal	Interest Rate	Time	Future Value	Compound Interest
a.	$13,000	12% compounded monthly	2 years	_____	_____
b.	$ 6,900	6% compounded quarterly	3 years	_____	_____

SOLUTION:

a. $0.12 \div 12 = 0.01$; $12 \times 2 = 24$; $\$13,000 \times 1.26973 = \$16,506.49$ Future Value
$\$16,506.49 - \$13,000 = \$3,506.49$ Compound Interest

b. $0.06 \div 4 = 0.015$; $4 \times 3 = 12$; $\$6,900 \times 1.19562 = \$8,249.78$ Future Value
$\$8,249.78 - \$6,900 = \$1,349.78$ Compound Interest

7.

Compute the future value for each of the following problems. Use Tables 16-1A&B or a calculator.

a. $4,650 is borrowed for 9 months at 9% compounded monthly   _____
b. $1,460 is loaned for 4 years at 8% compounded annually   _____

SOLUTION:

a. $0.09 \div 12 = 0.0075$; 9 months; $\$4,650 \times 1.06956 = \$4,973.45$
b. $0.08 \div 1 = 0.08$; $1 \times 4 = 4$; $\$1,460 \times 1.36049 = \$1,986.32$

8.

Compute the future value for each of the following problems. Use Tables 16-1A&B or a calculator.

a. $11,000 is borrowed for 2 1/2 years at 6% compounded quarterly   _____
b. $3,200 is invested for 3 years at 12% compounded semiannually   _____

SOLUTION:

a. $0.06 \div 4 = 0.015$; $4 \times 2\ 1/2 = 10$; $\$11,000 \times 1.16054 = \$12,765.94$
b. $0.12 \div 2 = 0.06$; $2 \times 3 = 6$; $\$3,200 \times 1.41852 = \$4,539.26$

9.

Compute the future value for each of the following problems. Use Tables 16-1A&B or a calculator.

a. $10,500 is deposited for 4 years at 10% compounded annually   _____
b. $800 is deposited for 11 years at 6% compounded semiannually   _____

SOLUTION:

a. $0.10 \div 1 = 0.10$; $1 \times 4 = 4$; $\$10,500 \times 1.46410 = \$15,373.05$
b. $0.06 \div 2 = 0.03$; $2 \times 11 = 22$; $\$800 \times 1.91610 = \$1,532.88$

10.
Compute the future value for each of the following problems. Use Tables
16-1A&B or a calculator.

a. $2,600 is loaned for 1 1/3 years at 15% compounded monthly   _____
b. $5,000 is invested for 3 years at 8% compounded quarterly   _____

SOLUTION:
a. $0.15 \div 12 = 0.0125$; $12 \times 1\ 1/3 = 16$; $\$2,600 \times 1.21989 = \$3,171.71$
b. $0.08 \div 4 = 0.02$; $4 \times 3 = 12$; $\$5,000 \times 1.26824 = \$6,341.20$

11.
University Lending Corp. loans money to students at 12% compounded semiannually. Dwight Burch
borrows $2,000 for 2 1/2 years. Compute the total that Dwight will be required to repay in both principal
and interest. (Use Tables 16-1A&B or a calculator.)

SOLUTION:
$0.12 \div 2 = 0.06$; $2 \times 2\ 1/2 = 5$
$\$2,000 \times 1.33823 = \$2,676.46$ Total Paid

12.
Keifer Air Conditioning can borrow money for 2 years at 12% compounded quarterly. Compute the
interest charge if Keifer borrows $28,000 to remodel their office. (Use Tables 16-1A&B or a calculator.)

SOLUTION:
$0.12 \div 4 = 0.03$; $4 \times 2 = 8$; $\$28,000 \times 1.26677 = \$35,469.56$
$\$35,469.56 - \$28,000 = \$7,469.56$ Interest

13.
Marie Chaney borrowed $2,460 from her mother for four years at 5% compounded annually. How much
interest will Marie have to pay to her mother? (Use Tables 16-1A&B or a calculator.)

SOLUTION:
$0.05 \div 1 = 0.05$; $1 \times 4 = 4$; $\$2,460 \times 1.21551 = \$2,990.15$
$\$2,990.15 - \$2,460 = \$530.15$ Interest

14.
Deborah Erickson deposited $1,625 in a credit union, which pays an interest of 8% compounded
quarterly. Compute the amount that Deborah will have in her account after 4 years. (Use Tables
16-1A&B or a calculator.)

SOLUTION:
$0.08 \div 4 = 0.02$; $4 \times 4 = 16$
$\$1,625 \times 1.37279 = \$2,230.78$ Account Balance

15.

Carol Chin received $5,400 as an annual profit-sharing bonus from her employer. Her accountant recommended an investment that would give her a return of 9% compounded annually. Compute the value of Carol's investment after 7 years. (Use Tables 16-1A&B or a calculator.)

SOLUTION:
0.09 ÷ 1 = 0.09; 1 × 7 = 7
$5,400 × 1.82804 = $9,871.42 value after 7 years

16.

When his daughter Sarah was born, Will Kravitz deposited $5,000 into an account that guaranteed to pay 8% compounded quarterly. Compute the amount that will be in the account six years later when Saraj starts school. (Use Tables 16-1A&B or a calculator.)

SOLUTION:
0.08 ÷ 4 = 0.02; 4 × 6 = 24
$5,000 × 1.60844 = $8,042.20 balance after 6 years

17.

Tracy Roberts loaned $3,600 to her sister Felicity, who just had a baby. Tracy charged her Felecity 6% compounded semiannually. Compute total payment that Tracy will receive from her sister if the loan is for 3 years. (Use Tables 16-1A&B or a calculator.)

SOLUTION:
0.06 ÷ 2 = 0.03; 2 × 3 = 6
$3,600 × 1.19405 = $4,298.58 total paid

18.

Winchell Plastics planned to buy a new delivery truck. They can borrow $16,000 for 15 months at 15% compounded monthly. If Winchell borrows the money at this rate, how much will they have to pay in interest? (Use Tables 16-1A&B or a calculator.)

SOLUTION:
0.15 ÷ 12 = 0.0125; 15 months; $16,000 × 1.20483 = $19,277.28
$19,277.28 − $16,000 = $3,277.28 interest

LEARNING OBJECTIVE 2, 3

19.

Compute the present value (principal) and the compound interest earned for each of the following investments. Use Tables 16-1A&B or 16-2A&B or a calculator.

	Future Value	Interest Rate	Time	Future Value	Compound Interest
a.	$ 6,880	5% compounded quarterly	7 years	_____	_____
b.	$12,500	5% compounded annually	13 years	_____	_____

SOLUTION:
a.   0.05 ÷ 4 = 0.0125; 4 × 7 = 28; $6,880 × 0.70622 = $4,858.79 Present Value
       $6,880 − $4,858.79 = $2,021.21 Compound Interest
b.   0.05 ÷ 1 = 0.05; 1 × 13 = 13; $12,500 × 0.53032 = $6,629.00 Present Value
       $12,500 − $6,629.00 = $5,871.00 Compound Interest

188   Part 4 INTEREST APPLICATIONS

20.
Compute the present value (principal) and the compound interest earned for each of the following
investments. Use Tables 16-1A&B or 16-2A&B or a calculator.

	Future Value	Interest Rate	Time	Future Value	Compound Interest
a.	$25,000	18% compounded monthly	3/4 year	_____	_____
b.	$ 2,650	6% compounded semiannually	10 years	_____	_____

SOLUTION:
a. $0.18 \div 12 = 0.015$; $12 \times 3/4 = 9$; $25,000 \times 0.87459 = $21,864.75$ Present Value
        $25,000 - $21,864.75 = $3,135.25$ Compound Interest

b. $0.06 \div 2 = 0.03$; $2 \times 10 = 20$; $2,650 \times 0.55368 = $1,467.25$ Present Value
        $2,650 - $1,467.25 = $1,182.75$ Compound Interest

21.
Compute the present value (principal) and the compound interest earned for each of the following
investments. Use Tables 16-1A&B or 16-2A&B or a calculator.

	Future Value	Interest Rate	Time	Future Value	Compound Interest
a.	$ 4,500	5% compounded annually	11 years	_____	_____
b.	$10,000	10% compounded semiannually	11 years	_____	_____

SOLUTION:
a. $0.05 \div 1 = 0.05$; $1 \times 11 = 11$; $4,500 \times 0.58468 = $2,631.06$ Present Value
        $4,500 - $2,631.06 = $1,868.94$ Compound Interest

b. $0.10 \div 2 = 0.05$; $2 \times 11 = 22$; $10,000 \times 0.34185 = $3,418.50$
        $10,000 - $3,418.50 = $6,581.50$

22.
Compute the present value (principal) and the compound interest earned for each of the following
investments. Use Tables 16-1A&B or 16-2A&B or a calculator.

	Future Value	Interest Rate	Time	Future Value	Compound Interest
a.	$1,700	9% compounded monthly	10 months	_____	_____
b.	$6,250	6% compounded quarterly	6 years	_____	_____

SOLUTION:
a. $0.09 \div 12 = 0.0075$; 10 months; $1,700 \times 0.92800 = $1,577.60$ Present Value
        $1,700 - $1,577.60 = 122.40 Compound Interest

b. $0.06 \div 4 = 0.015$; $4 \times 6 = 24$; $6,250 \times 0.69954 = $4,372.13$ Present Value
        $6,250 - $4,372.13 = $1,877.87$ Compound Interest

23.
Compute the present value (principal) and the compound interest earned for each of the following investments. Use Tables 16-1A&B or 16-2A&B or a calculator.

	Future Value	Interest Rate	Time	Future Value	Compound Interest
a.	$15,750	8% compounded annually	18 years	_____	_____
b.	$ 4,280	12% compounded monthly	2 1/2 years	_____	_____

SOLUTION:
a. $0.08 \div 1 = 0.08$; $1 \times 18 = 18$; $15,750 \times 0.25025 = \$3,941.44$ Present Value
     $15,750 - \$3,941.44 = \$11,808.56$ Compound Interest

b. $0.12 \div 12 = 0.01$; $12 \times 2\ 1/2 = 30$; $4,280 \times 0.74192 = \$3,175.42$ Present Value
     $4,280 - \$3,175.42 = \$1,104.58$ Compound Interest

24.
Compute the present value (principal) and the compound interest earned for each of the following investments. Use Tables 16-1A&B or 16-2A&B or a calculator.

	Future Value	Interest Rate	Time	Future Value	Compound Interest
a.	$5,240	8% compounded quarterly	7 years	_____	_____
b.	$7,110	10% compounded semiannually	10 years	_____	_____

SOLUTION:
a. $0.08 \div 4 = 0.02$; $4 \times 7 = 28$; $5,240 \times 0.57437 = \$3,009.70$ Present Value
     $5,240 - \$3,009.70 = \$2,230.30$ Compound Interest

b. $0.10 \div 2 = 0.05$; $2 \times 10 = 20$; $7,110 \times 0.37689 = \$2,679.69$ Present Value
     $7,110 - \$2,679.69 = \$4,430.31$ Compound Interest

25.
Compute the present value (principal) and the compound interest earned for each of the following investments. Use Tables 16-1A&B or 16-2A&B or a calculator.

	Future Value	Interest Rate	Time	Future Value	Compound Interest
a.	$8,000	12% compounded semiannually	14 years	_____	_____
b.	$3,675	4% compounded quarterly	9 years	_____	_____

SOLUTION:
a. $0.12 \div 2 = 0.06$; $2 \times 14 = 28$; $8,000 \times 0.19563 = \$1,565.04$ Present Value
     $8,000 - \$1,565.04 = \$6,434.96$ Compound Interest

b. $0.04 \div 4 = 0.01$; $4 \times 9 = 36$; $3,675 \times 0.69892 = \$2,568.53$ Present Value
     $3,675 - \$2,568.53 = \$1,106.47$ Compound Interest

26.

Compute the present value (principal) and the compound interest earned for each of the following investments. Use Tables 16-1A&B or 16-2A&B or a calculator.

	Future Value	Interest Rate	Time	Future Value	Compound Interest
a.	$10,085	8% compounded annually	11 years	_____	_____
b.	$ 6,000	6% compounded monthly	4 years	_____	_____

SOLUTION:

a. $0.08 \div 1 = 0.08$; $1 \times 11 = 11$; $10,085 \times 0.42888 = $4,325.25$ Present Value
        $10,085 - $4,325.25 = $5,759.75$ Compound Interest

b. $0.06 \div 12 = 0.005$; $12 \times 4 = 48$; $6,000 \times 0.78710 = $4,722.60$ Present Value
        $6,000 - $4,722.60 = $1,277.40$ Compound Interest

27.

Compute the present value in each of the following problems. Use Tables 16-1A&B or 16-2A&B or a calculator.

a.   Compute the amount that you must lend today at 10% compounded semiannually to be repaid a total (principal and interest) of $10,000 in 13 years.

b.   Compute the amount that you must invest today at 12% compounded annually to have $1,500 in 3 years.

SOLUTION:

a. $0.10 \div 2 = 0.05$; $2 \times 13 = 26$; $10,000 \times 0.28124 = $2,812.40$ Lend today
b. $0.12 \div 1 = 0.12$; $1 \times 3 = 3$; $1,500 \times 0.71178 = $1,067.67$ Invest today

28.

Compute the present value in each of the following problems. Use Tables 16-1A&B or 16-2A&B or a calculator.

a.   How much must you deposit today into an account that pays 12% compounded quarterly to have $9,000 in 3 years?

b.   How much must you lend today at 12% compounded semiannually to be repaid a total (principal and interest) of $3,250 in 5 years?

SOLUTION:

a. $0.12 \div 4 = 0.03$; $4 \times 3 = 12$; $9,000 \times 0.70138 = $6,312.42$
b. $0.12 \div 2 = 0.06$; $2 \times 5 = 10$; $3,250 \times 0.55839 = $1,814.77$

29.
Compute the present value in each of the following problems. Use Tables 16-1A&B or 16-2A&B or a calculator.

a.   How much must you invest today at 12% compounded monthly to have $10,000 in 2 years?

b.   Compute the amount that you must deposit today into an account that pays 6% compounded quarterly to have $4,000 in 4 years.

SOLUTION:
a. $0.12 \div 12 = 0.01$; $12 \times 2 = 24$; $\$10,000 \times 0.78757 = \$7,875.70$

b. $0.06 \div 4 = 0.015$; $4 \times 4 = 16$; $\$4,000 \times 0.78803 = \$3,152.12$ Deposit today

LEARNING OBJECTIVES 1, 2, 3

30.
Harold Lau will deposit enough money today so that his account will contain $20,000 in ten years. The account will pay interest at 8% compounded semiannually. Compute the interest (in dollars) that Harold will earn during the ten years. (Use Tables 16-1A&B or 16-2A&B or a calculator.)

SOLUTION:
$0.08 \div 2 = 0.04$; $2 \times 10 = 20$; $\$20,000 \times 0.45639 = \$9,127.80$
$\$20,000 - \$9,127.80 = \$10,872.20$ Interest

31.
Beverly Forest is a single parent with twin daughters who are now 16 years old. Beverly inherited money from her father's estate and decided that it would be reasonable to save $12,000 of the inheritance for wedding expenses in the event that both daughters should decide to get married in the same year. How much money will Beverly have in 6 years if she can invest the $12,000 and get a return of 12% compounded quarterly? (Use Tables 16-1A&B or 16-2A&B or a calculator.)

SOLUTION:
$0.12 \div 4 = 0.03$; $4 \times 6 = 24$
$\$12,000 \times 2.03279 = \$24,393.48$ in 6 years

32.
Newton Kress plans to give his son $5,000 when he is 21, which will be 5 years from now. If Newton finds an investment that will pay 8% compounded quarterly, compute the amount that he must invest today to achieve his goal. (Use Tables 16-1A&B or 16-2A&B or a calculator.)

SOLUTION:
$0.08 \div 4 = 0.02$; $4 \times 5 = 20$
$\$5,000 \times 0.67297 = \$3,364.85$ Invest today

33.

Two-and-one-half years from now, Melanie Olson wants to have $3,500 in the bank. She can earn interest of 12% compound monthly. Compute the amount that Melanie must deposit today. (Use Tables 16-1A&B or 16-2A&B or a calculator.)

SOLUTION:
0.12 ÷ 12 = 0.01; 12 × 2 1/2 = 30
$3,500 × 0.74192 = $2,596.72 Deposit today

34.

Amanda Nelson wants to buy a new car 5 years from now. She estimates that she will need $12,000. Compute the amount that Amanda must invest if she can earn 6% compounded quarterly. (Use Tables 16-1A&B or 16-2A&B or a calculator.)

SOLUTION:
0.06 ÷ 4 = 0.015; 4 × 5 = 20
$12,000 × 0.74247 = $8,909.64 Invest now

35.

Maria Gomez, an attorney, plans to replace her office furniture in three years. She estimates the cost will be $7,000. Compute the amount that Maria should deposit today at 6% compounded annually to have the money available. (Use Tables 16-1A&B or 16-2A&B or a calculator.)

SOLUTION:
0.06 ÷ 1 = 0.06; 1 × 3 = 3
$7,000 × 0.83962 = $5,877.34 Deposit today

36.

When she got married, Lannie Ferguson left all of her personal savings in her own credit union account, which was paying 8% compounded quarterly. Five years later, the same account had increased to $6,425.84. Compute the amount that was in the account when Lannie got married. She made no additional deposits into the account.   (Use Tables 16-1A&B or 16-2A&B or a calculator.)

SOLUTION:
0.08 ÷ 4 = 0.02; 4 × 5 = 20
$6,425.84 × 0.67297 = $4,324.40 in account

37.

In January, Dana Blakely decided to donate $2,500 to the Riverfront Humane Society. For income tax purposes she will not make the donation until next December. If Dana can earn 12% compounded monthly, how much must she invest in January to have $2,500 in 11 months? (Use Tables 16-1A&B or 16-2A&B or a calculator.)

SOLUTION:
0.12 ÷ 12 = 0.01; 11 months
$2,500 × 0.89632 = $2,240.80 Invest in January

Period (n)	0.50%	0.75%	1.00%	1.25%	1.50%	2.00%	3.00%
1	1.00500	1.00750	1.01000	1.01250	1.01500	1.02000	1.03000
2	1.01002	1.01506	1.02010	1.02516	1.03022	1.04040	1.06090
3	1.01508	1.02267	1.03030	1.03797	1.04568	1.06121	1.09273
4	1.02015	1.03034	1.04060	1.05095	1.06136	1.08243	1.12551
5	1.02525	1.03807	1.05101	1.06408	1.07728	1.10408	1.15927
6	1.03038	1.04585	1.06152	1.07738	1.09344	1.12616	1.19405
7	1.03553	1.05370	1.07214	1.09085	1.10984	1.14869	1.22987
8	1.04071	1.06160	1.08286	1.10449	1.12649	1.17166	1.26677
9	1.04591	1.06956	1.09369	1.11829	1.14339	1.19509	1.30477
10	1.05114	1.07758	1.10462	1.13227	1.16054	1.21899	1.34392
11	1.05640	1.08566	1.11567	1.14642	1.17795	1.24337	1.38423
12	1.06168	1.09381	1.12683	1.16075	1.19562	1.26824	1.42576
13	1.06699	1.10201	1.13809	1.17526	1.21355	1.29361	1.46853
14	1.07232	1.11028	1.14947	1.18995	1.23176	1.31948	1.51259
15	1.07768	1.11860	1.16097	1.20483	1.25023	1.34587	1.55797
16	1.08307	1.12699	1.17258	1.21989	1.26899	1.37279	1.60471
17	1.08849	1.13544	1.18430	1.23514	1.28802	1.40024	1.65285
18	1.09393	1.14396	1.19615	1.25058	1.30734	1.42825	1.70243
19	1.09940	1.15254	1.20811	1.26621	1.32695	1.45681	1.75351
20	1.10490	1.16118	1.22019	1.28204	1.34686	1.48595	1.80611
21	1.11042	1.16989	1.23239	1.29806	1.36706	1.51567	1.86029
22	1.11597	1.17867	1.24472	1.31429	1.38756	1.54598	1.91610
23	1.12155	1.18751	1.25716	1.33072	1.40838	1.57690	1.97359
24	1.12716	1.19641	1.26973	1.34735	1.42950	1.60844	2.03279
25	1.13280	1.20539	1.28243	1.36419	1.45095	1.64061	2.09378
26	1.13846	1.21443	1.29526	1.38125	1.47271	1.67342	2.15659
27	1.14415	1.22354	1.30821	1.39851	1.49480	1.70689	2.22129
28	1.14987	1.23271	1.32129	1.41599	1.51722	1.74102	2.28793
29	1.15562	1.24196	1.33450	1.43369	1.53998	1.77584	2.35657
30	1.16140	1.25127	1.34785	1.45161	1.56308	1.81136	2.42726

Table 16-1A. Future Value (Compound) Factors

Period (n)	4.00%	5.00%	6.00%	8.00%	9.00%	10.00%	12.00%
1	1.04000	1.05000	1.06000	1.08000	1.09000	1.10000	1.12000
2	1.08160	1.10250	1.12360	1.16640	1.18810	1.21000	1.25440
3	1.12486	1.15763	1.19102	1.25971	1.29503	1.33100	1.40493
4	1.16986	1.21551	1.26248	1.36049	1.41158	1.46410	1.57352
5	1.21665	1.27628	1.33823	1.46933	1.53862	1.61051	1.76234
6	1.26532	1.34010	1.41852	1.58687	1.67710	1.77156	1.97382
7	1.31593	1.40710	1.50363	1.71382	1.82804	1.94872	2.21068
8	1.36857	1.47746	1.59385	1.85093	1.99256	2.14359	2.47596
9	1.42331	1.55133	1.68948	1.99900	2.17189	2.35795	2.77308
10	1.48024	1.62889	1.79085	2.15892	2.36736	2.59374	3.10585
11	1.53945	1.71034	1.89830	2.33164	2.58043	2.85312	3.47855
12	1.60103	1.79586	2.01220	2.51817	2.81266	3.13843	3.89598
13	1.66507	1.88565	2.13293	2.71962	3.06580	3.45227	4.36349
14	1.73168	1.97993	2.26090	2.93719	3.34173	3.79750	4.88711
15	1.80094	2.07893	2.39656	3.17217	3.64248	4.17725	5.47357
16	1.87298	2.18287	2.54035	3.42594	3.97031	4.59497	6.13039
17	1.94790	2.29202	2.69277	3.70002	4.32763	5.05447	6.86604
18	2.02582	2.40662	2.85434	3.99602	4.71712	5.55992	7.68997
19	2.10685	2.52695	3.02560	4.31570	5.14166	6.11591	8.61276
20	2.19112	2.65330	3.20714	4.66096	5.60441	6.72750	9.64629
21	2.27877	2.78596	3.39956	5.03383	6.10881	7.40025	10.80385
22	2.36992	2.92526	3.60354	5.43654	6.65860	8.14027	12.10031
23	2.46472	3.07152	3.81975	5.87146	7.25787	8.95430	13.55235
24	2.56330	3.22510	4.04893	6.34118	7.91108	9.84973	15.17863
25	2.66584	3.38635	4.29187	6.84848	8.62308	10.83471	17.00006
26	2.77247	3.55567	4.54938	7.39635	9.39916	11.91818	19.04007
27	2.88337	3.73346	4.82235	7.98806	10.24508	13.10999	21.32488
28	2.99870	3.92013	5.11169	8.62711	11.16714	14.42099	23.88387
29	3.11865	4.11614	5.41839	9.31727	12.17218	15.86309	26.74993
30	3.24340	4.32194	5.74349	10.06266	13.26768	17.44940	29.95992

Table 16-1B. Future Value (Compound) Factors

Period (n)	0.50%	0.75%	1.00%	1.25%	1.50%	2.00%	3.00%
1	0.99502	0.99256	0.99010	0.98765	0.98522	0.98039	0.97087
2	0.99007	0.98517	0.98030	0.97546	0.97066	0.96117	0.94260
3	0.98515	0.97783	0.97059	0.96342	0.95632	0.94232	0.91514
4	0.98025	0.97055	0.96098	0.95152	0.94218	0.92385	0.88849
5	0.97537	0.96333	0.95147	0.93978	0.92826	0.90573	0.86261
6	0.97052	0.95616	0.94205	0.92817	0.91454	0.88797	0.83748
7	0.96569	0.94904	0.93272	0.91672	0.90103	0.87056	0.81309
8	0.96089	0.94198	0.92348	0.90540	0.88771	0.85349	0.78941
9	0.95610	0.93496	0.91434	0.89422	0.87459	0.83676	0.76642
10	0.95135	0.92800	0.90529	0.88318	0.86167	0.82035	0.74409
11	0.94661	0.92109	0.89632	0.87228	0.84893	0.80426	0.72242
12	0.94191	0.91424	0.88745	0.86151	0.83639	0.78849	0.70138
13	0.93722	0.90743	0.87866	0.85087	0.82403	0.77303	0.68095
14	0.93256	0.90068	0.86996	0.84037	0.81185	0.75788	0.66112
15	0.92792	0.89397	0.86135	0.82999	0.79985	0.74301	0.64186
16	0.92330	0.88732	0.85282	0.81975	0.78803	0.72845	0.62317
17	0.91871	0.88071	0.84438	0.80963	0.77639	0.71416	0.60502
18	0.91414	0.87416	0.83602	0.79963	0.76491	0.70016	0.58739
19	0.90959	0.86765	0.82774	0.78976	0.75361	0.68643	0.57029
20	0.90506	0.86119	0.81954	0.78001	0.74247	0.67297	0.55368
21	0.90056	0.85478	0.81143	0.77038	0.73150	0.65978	0.53755
22	0.89608	0.84842	0.80340	0.76087	0.72069	0.64684	0.52189
23	0.89162	0.84210	0.79544	0.75147	0.71004	0.63416	0.50669
24	0.88719	0.83583	0.78757	0.74220	0.69954	0.62172	0.49193
25	0.88277	0.82961	0.77977	0.73303	0.68921	0.60953	0.47761
26	0.87838	0.82343	0.77205	0.72398	0.67902	0.59758	0.46369
27	0.87401	0.81730	0.76440	0.71505	0.66899	0.58586	0.45019
28	0.86966	0.81122	0.75684	0.70622	0.65910	0.57437	0.43708
29	0.86533	0.80518	0.74934	0.69750	0.64936	0.56311	0.42435
30	0.86103	0.79919	0.74192	0.68889	0.63976	0.55207	0.41199

Table 16-2A. Present Value Factors

Period (n)	4.00%	5.00%	6.00%	8.00%	9.00%	10.00%	12.00%
1	0.96154	0.95238	0.94340	0.92593	0.91743	0.90909	0.89286
2	0.92456	0.90703	0.89000	0.85734	0.84168	0.82645	0.79719
3	0.88900	0.83684	0.83962	0.79383	0.77218	0.75131	0.71178
4	0.85480	0.82270	0.79209	0.73503	0.70843	0.68301	0.63552
5	0.82193	0.78353	0.74726	0.68058	0.64993	0.62092	0.56743
6	0.79031	0.74622	0.70496	0.63017	0.59627	0.56447	0.50663
7	0.75992	0.71068	0.66506	0.58349	0.54703	0.51316	0.45235
8	0.73069	0.67684	0.62741	0.54027	0.50187	0.46651	0.40388
9	0.70259	0.64461	0.59190	0.50025	0.46043	0.42410	0.36061
10	0.67556	0.61391	0.55839	0.46319	0.42241	0.38554	0.32197
11	0.64958	0.58468	0.52679	0.42888	0.38753	0.35049	0.28748
12	0.62460	0.55684	0.49697	0.39711	0.35553	0.31853	0.25668
13	0.60057	0.53032	0.46884	0.36770	0.32618	0.28966	0.22917
14	0.57748	0.50507	0.44230	0.34046	0.29925	0.26333	0.20462
15	0.55526	0.48102	0.41727	0.31524	0.27454	0.23939	0.18270
16	0.53391	0.45811	0.39365	0.29189	0.25187	0.21763	0.16312
17	0.51337	0.43630	0.37136	0.27027	0.23107	0.19784	0.14564
18	0.49363	0.41552	0.35034	0.25025	0.21199	0.17986	0.13004
19	0.47464	0.39573	0.33051	0.23171	0.19449	0.16351	0.11611
20	0.45639	0.37689	0.31180	0.21455	0.17843	0.14864	0.10367
21	0.43883	0.35894	0.29416	0.19866	0.16370	0.13513	0.09256
22	0.42196	0.34185	0.27751	0.18394	0.15018	0.12285	0.08264
23	0.40573	0.32557	0.26180	0.17032	0.13778	0.11168	0.07379
24	0.39012	0.31007	0.24698	0.15770	0.12640	0.10153	0.06588
25	0.37512	0.29530	0.23300	0.14602	0.11597	0.09230	0.05882
26	0.36069	0.28124	0.21981	0.13520	0.10639	0.08391	0.05252
27	0.34682	0.26785	0.20737	0.12519	0.09761	0.07628	0.04689
28	0.33348	0.25509	0.19563	0.11591	0.08955	0.06934	0.04187
29	0.32065	0.24295	0.18456	0.10733	0.08215	0.06304	0.03738
30	0.30832	0.23138	0.17411	0.09938	0.07537	0.05731	0.03338

Table 16-2B. Present Value Factors

Chapter 17   INVENTORY AND TURNOVER

PROBLEMS

LEARNING OBJECTIVE 1

1.
Walker's Shoe Stores shows an inventory of 372 pairs of shoes at a price of $37.28 per pair. Compute the amount that should be shown in the extension column.

SOLUTION: $13,868.16

2.
John's Tires ended the month with 420 tires. There were 300 of Brand A at a price of $47.50 each and 120 of Brand B at a price of $51.00 each. Compute the amount of inventory at month end.

SOLUTION: 20,370 $(300 \times 47.5) + (120 \times 51)$

3.
The Majestic Theater had 378 cans of Kettle Corn and 132 cans of regular Pop Corn on hand at month end. During the next month Majestic sold 142 cans of Kettle Corn and 67 cans of regular Pop Corn. Cans of each cost $28.40. What is the value of the inventory on hand at the end of the second month?

SOLUTION: $8,548.40

4.
An inventory record sheet shows a balance on hand to be 382. After units out of 102 and 81, compute the amount that should be shown in the balance on hand column.

SOLUTION: 199

5.
An inventory record sheet shows a balance on hand of 972. After units out of 111, 204, and 87 and units in of 700 what is the amount remaining?

SOLUTION: 1,270

6.
Mike Hatcher Ski Shop had 309 pairs of skis in stock at the beginning of the month.   In the four weeks of the month the shop sold respectfully 32, 19, 46, and 22 pairs of skis. During the month Mike bought 39 additional pairs of skis for his inventory.   What was the month end inventory?

SOLUTIION: 229

7.
An inventory record sheet shows a balance on hand to be 260. After units in of 97 and 88, compute the amount that should be shown in the balance on hand column.

SOLUTION: 445

8.

Fran's Fourth Street Flower Shop showed an inventory balance of 431 plants at the end of February. During March Fran sold 70 plants and brought in 39 plants. During April Fran sold 106 plants and brought in 111 plants.   What was the inventory for Fran's Fourth Street Flower Shop at the end of April?

SOLUTION: 405

LEARNING OBJECTIVE 2

9.

A small-appliance merchant using the average cost method of valuing inventory has 65 toasters remaining in inventory. The merchant purchased toasters over a three-month period as follows: 40 purchased at $8.00 on April 1, 60 purchased at $7.70 on May 1, and 20 purchased at $7.80 on June 1. Compute the value of the ending inventory of toasters at average cost.

SOLUTION: $508.08   ($938 / 120 = $7.82) 65 × 7.82 = 508.30 or ($938 / 120 = $7.81666 × 65 = $508.08

10.

A small-appliance merchant using the FIFO method of valuing inventory has 60 toasters remaining in inventory. The merchant purchased toasters over a three month-period as follows: 30 purchased at $7.80 on April 1, 50 purchased at $7.70 on May 1, and 20 purchased at $7.80 on June 1. Compute the value of the ending inventory of toasters at FIFO cost.

SOLUTION: $464

11.

A small-appliance merchant using the LIFO method of valuing inventory has 60 toasters remaining in inventory. The merchant purchased toasters over a three-month period as follows: 30 purchased at $7.60 on April 1, 50 purchased at $7.70 on May 1, and 20 purchased at $7.80 on June 1. Compute the value of the ending inventory of toasters at LIFO cost.

SOLUTION: $459

12.

A hardware merchant using the average cost method of valuing inventory has 350 hammers remaining in inventory. The merchant purchased hammers over a three-month period as follows: 120 purchased at $3.50 in April, 150 purchased at $3.62 in May, and 150 purchased at $3.72 in June. Compute the value of the ending inventory of hammers at average cost.

SOLUTION: $1,267.50

13.

A hardware merchant using the FIFO method of valuing inventory has 300 hammers remaining in inventory. The merchant purchased hammers over a three month-period as follows: 120 purchased at $3.50 on April 1, 150 purchased at $3.62 on May 1, and 150 purchased at $3.72 on June 1. Compute the value of the ending inventory of hammers at FIFO cost.

SOLUTION: $1,101

14.
A hardware merchant using the LIFO method of valuing inventory has 200 hammers remaining in inventory. The merchant purchased hammers over a three-month period as follows: 120 purchased at $3.50 on April 1, 150 purchased at $3.62 on May 1, and 150 purchased at $3.72 on June 1. Compute the value of the ending inventory of hammers at LIFO cost.

SOLUTION: $709.60

15.
Hammond Stationery uses the LIFO method of valuing inventory and has 350 pen sets remaining in inventory. Hammond purchased the pen sets over a six month period as follows: May 1: 175 at $20.50; May 15: 125 at $22.00; July 7: 75 at $24.00; and October 1: 150 at $23.50. Compute the value of the ending inventory

SOLUTION: $7,537.50

16.
Culver Clocks, Inc. uses the LIFO method of valuing inventory. It purchased clocks as follows over the past year: January 1: 100 at $38.000; March 1: 250 at $36.00; July 1: 175 at $39.00; October 1: 400 at $38.50; December 1: 300 at $39.50.   At year end on December 31, Culver Clocks, Inc. had a year-end inventory of 300 clocks. Compute the December 31, inventory value.

SOLUTION: $11,000.

17.
Special Sports Store, which uses the average cost method of valuing inventory, has 52 tennis rackets remaining in inventory. Special purchased tennis rackets over a three month-period as follows: 36 purchased at $15 on January 3, 48 purchased at $15.50 on February 5, and 24 purchased at $16 on March 7. Compute the value of the ending inventory of tennis rackets at average cost.

SOLUTION: $803.11

18.
Special Sports Store, which uses the FIFO method of valuing inventory, has 60 tennis rackets remaining in inventory. Special purchased tennis rackets over a three-month period as follows: 36 purchased at $15 on January 3, 48 purchased at $15.50 on February 5, and 24 purchased at $16 on March 7. Compute the value of the ending inventory of tennis rackets at FIFO cost.

SOLUTION: $942

19.
Special Sports Store, which uses the LIFO method of valuing inventory, has 52 tennis rackets remaining in inventory. Special purchased tennis rackets over a three-month period as follows: 36 purchased at $15 on January 3, 48 purchased at $15.50 on February 5, and 24 purchased at $16 on March 7. Compute the value of the ending inventory of tennis rackets at LIFO cost.

SOLUTION: $788

20.

Toddlers' Toyland using the average cost method of valuing inventory has 75 brown bears remaining in inventory. Toddlers' purchased brown bears over a three-month period as follows: 72 purchased at $3.50 on October 7; 84 purchased at $3.65 on November 7; and 60 purchased at $3.80 on December 7. Compute the value of the ending inventory of bears at average cost.

SOLUTION: $273.13

21.

Toddlers' Toyland, which uses the FIFO method of valuing inventory, has 75 brown bears remaining in inventory. Toddlers' purchased brown bears over a three-month period as follows: 72 purchased at $3.50 on October 7, 84 purchased at $3.65 on November 7, and 60 purchased at $3.80 on December 7. Compute the value of the ending inventory of bears at FIFO cost.

SOLUTION: $282.75

22.

Toddlers' Toyland, which uses the LIFO method of valuing inventory, has 80 brown bears remaining in inventory. Toddlers' purchased brown bears over a three-month period as follows: 72 purchased at $3.50 on October 7, 84 purchased at $3.65 on November 7, and 60 purchased at $3.80 on December 7. Compute the value of the ending inventory of bears at LIFO cost.

SOLUTION: $281.20

23.

Home and Hearth, which uses the average cost method of valuing inventory, has 35 fire screens remaining in inventory. Home and Hearth purchased fire screens over a 12-month period as follows: 36 purchased at $29 on January 4, 18 purchased at $27 on April 2, 18 purchased at $25 on July 30; and 36 purchased at $23 on September 1. Compute the value of the ending inventory of fire screens at average cost.

SOLUTION: $910

24.

Home and Hearth, which uses the FIFO method of valuing inventory, has 50 fire screens remaining in inventory. Home and Hearth purchased fire screens over a 12-month period as follows: 36 purchased at $29 on January 4, 18 purchased at $27 on April 2, 18 purchased at $25 on July 30, and 36 purchased at $23 on September 1. Compute the value of the ending inventory of fire screens at FIFO cost.

SOLUTION: $1,178

25.

Home and Hearth, which uses the LIFO method of valuing inventory, has 35 fire screens remaining in inventory. Home and Hearth purchased fire screens over a 12-month period as follows: 36 purchased at $29 on January 4, 18 purchased at $27 on April 2, 18 purchased at $25 on July 30, and 36 purchased at $23 on September 1. Compute the value of the ending inventory of fire screens at LIFO cost.

SOLUTION: $1,015

LEARNING OBJECTIVE 3

26.
A small-appliance merchant shows an inventory of 200 toasters at a cost of $7.25 and a market value of $8.00. Compute the inventory value at the lower of cost or market.

SOLUTION: $1,450

27.
A small-appliance merchant shows an inventory of 150 waffle irons at a cost of $7.80 and a market value of $7.40. Compute the inventory value at the lower of cost or market.

SOLUTION: $1,110

28.
Dalton Boats, Inc. counted a year end inventory of 78 boats as follows:

Number of Boats	Cost per Unit	Retail Value per Unit
18	$28,800	$37,400
20	$36,400	$33,900
30	$18,700	$23,050
10	$52,000	$51,000

Compute the inventory value at the lower of cost or market.

SOLUTION: $2,267,400

29.
Dalton Boats, Inc. counted an inventory at the end of January of 22 boats as follows:

Number of Boats	Cost per Unit	Retail Value per Unit
8	$28,800	$42,000
2	$36,400	$39,900
9	$18,700	$23,050
3	$26,000	$24,500

Compute the inventory value at the lower of cost or market.

SOLUTION: $545,000

30.
A small-appliance merchant shows an inventory of electric items with costs and retail values as follows:

Item	Number	Cost	Retail Value
Electric Toaster	20	$29.00	$34.00
Coffee Makers	30	$31.00	$30.50
Microwave Ovens	8	$99.00	$122.00

Compute the inventory value at the lower of cost or market.

SOLUTION: $2,287.00

31.
A small-appliance merchant shows an inventory of 63 blenders at a cost of $9.65 and a market value of $9.30. Compute the inventory value at the lower of cost or market.

SOLUTION: $585.90

LEARNING OBJECTIVE 4

32.

Willis Hardware had a beginning inventory of $160,000 at cost. During the month, Willis purchased and received $100,000 in goods and had net sales of $200,000. Throughout the month, Willis maintained a 40% markup on all sales. What was the cost of goods sold?

SOLUTION: $120,000

33.

Willis Hardware had a beginning inventory of $160,000 at cost. During the month, Willis purchased and received $100,000 in goods and had net sales of $200,000. Throughout the month, Willis maintained a 40% markup on all sales. Compute the ending inventory at cost.

SOLUTION: $140,000

34.

Brammer Appliance had a beginning inventory of $40,000 at cost. During the month, Brammer purchased and received $25,000 in goods and had net sales of $50,000. Throughout the month, Brammer maintained a 30% markup on all sales. Compute the cost of goods sold.

SOLUTION: $35,000

35.

Brammer Appliance had a beginning inventory of $40,000 at cost. During the month, Brammer purchased and received $25,000 in goods and had net sales of $50,000. Throughout the month, Brammer maintained a 30% markup on all sales. Compute the ending inventory at cost.

SOLUTION: $30,000

36.

Bright Pots and Pans had a beginning inventory of $240,000 at cost. During the month, Bright purchased and received $150,000 in goods and had net sales of $280,000. Throughout the month, Bright maintained a 50% markup on all sales. Compute the cost of goods sold.

SOLUTION: $140,000

37.

Bright Pots and Pans had a beginning inventory of $240,000 at cost. During the month, Bright purchased and received $150,000 in goods and had net sales of $280,000. Throughout the month, Bright maintained a 50% markup on all sales. Compute the ending inventory at cost.

SOLUTION: $250,000

38.

Finished Furniture had a beginning inventory of $110,000 at cost. During the month, Finished purchased and received $60,000 in goods and had net sales of $150,000. Throughout the month, Finished Furniture maintained a 40% markup on all sales. Compute the cost of goods sold.

SOLUTION: $90,000

39.

Finished Furniture had a beginning inventory of $110,000 at cost. During the month, Finished purchased and received $60,000 in goods and had net sales of $150,000. Throughout the month, Finished Furniture maintained a 40% markup on all sales. Compute the ending inventory at cost.

SOLUTION: $80,000

40.

Happy Heart Exercise Equipment Company had a beginning inventory of $160,000 at cost. During the month, Happy purchased and received $120,000 in goods and had net sales of $130,000. Throughout the month, Happy Heart Exercise Equipment Company maintained a 35% markup on all sales. Compute the cost of goods sold.

SOLUTION: $84,500

41.

Happy Heart Exercise Equipment Company had a beginning inventory of $160,000 at cost. During the month, Happy purchased and received $120,000 in goods and had net sales of $130,000. Throughout the month, Happy Heart Exercise Equipment Company maintained a 35% markup on all sales. Compute the ending inventory at cost.

SOLUTION: $195,500

LEARNING OBJECTIVE 5

42.

Curtis Auto Parts takes inventory annually. Beginning inventory for the year was $196,000. Ending inventory for the year was $212,000. Compute the average inventory for Curtis Auto Parts.

SOLUTION: $204,000

43.

Pretty Paintings Gallery takes inventory semi-annually. Beginning inventory for the year was $284,000. Inventory at the end of the first six months was $256,000. Ending inventory for the year was $240,000. Compute the average inventory for Pretty Paintings Gallery.

SOLUTION: $260,000

44.

Fun and Frolic Sporting Goods takes inventory quarterly. Beginning inventory for the year was $398,000. Inventory at the end of the first quarter was $408,000; at the end of the second quarter, $372,000; at the end of the third quarter, $360,000; and at the end of the year, $412,000. Compute the average inventory for Fun and Frolic Sporting Goods.

SOLUTION: $390,000

45.

Sharp's Cutlery takes inventory annually. Beginning inventory for the year was $89,000. Ending inventory for the year was $80,000. Compute the average inventory for Sharp's Cutlery.

SOLUTION: $84,500

46.

Crazy Quilts takes inventory semi-annually. Beginning inventory for the year was $32,000. Inventory at the end of the first six months was $27,000. Ending inventory for the year was $39,000. Compute the average inventory for Crazy Quilts.

SOLUTION: $32,666.67

47.

Rainy Day Umbrellas takes inventory quarterly. Beginning inventory for the year was $18,200. Inventory at the end of the first quarter was $15,400; at the end of the second quarter, $13,600; at the end of the third quarter, $17,500; and at the end of the year, $19,300. Compute the average inventory for Rainy Day Umbrellas.

SOLUTION: $16,800

48.

Readers' Bookstore takes inventory annually and bases inventory on selling price. Beginning inventory for the year was $44,150. Ending inventory for the year was $39,350. Net sales for the year equal $75,150. Compute the inventory turnover at retail.

SOLUTION: 1.8

49.

Purrfect Pet Shop takes inventory semi-annually and bases inventory on selling price. Beginning inventory for the year was $67,320. Inventory at the end of the first six months was $79,680. Ending inventory for the year was $82,500. Net sales for the year equal $183,750. Compute the inventory turnover at retail.

SOLUTION: 2.4

50.

Good Grooming Supplies takes inventory quarterly and bases inventory on selling price. Beginning inventory for the year was $91,000. Inventory at the end of the first quarter was $89,500; at the end of the second quarter, $74,120; at the end of the third quarter, $82,210; and at the end of the year, $79,170. Net sales for the year equal $249,600. Compute the inventory turnover at retail.

SOLUTION: 3

51.

Junk Jewelry, Inc. takes inventory annually and bases inventory on selling price. Beginning inventory for the year was $38,600. Ending inventory for the year was $31,600. Net sales for the year equal $97,600. Compute the inventory turnover at retail.

SOLUTION: 2.78 or 2.8

52.
Favorite Flavors, Inc. takes inventory semi-annually and bases inventory on selling price. Beginning inventory for the year was $124,000. Inventory at the end of the first six months was $112,000. Ending inventory for the year was $133,000. Net sales for the year equal $430,500. Compute the inventory turnover at retail.

SOLUTION: 3.5

53.
Bud's Blooms takes inventory quarterly and bases inventory on selling price. Beginning inventory for the year was $64,500. Inventory at the end of the first quarter was $67,200; at the end of the second quarter, $71,000; at the end of the third quarter, $73,000; and at the end of the year, $65,300. Net sales for the year equal $190,960. Compute the inventory turnover at retail.

SOLUTION: 2.8

54.
A merchant keeping inventory on cost price had a beginning inventory of $30,000, purchases of $150,000, and an ending inventory of $40,000. Compute the cost of goods sold.

SOLUTION: $140,000

55.
A merchant had a beginning inventory of $30,000, purchases of $150,000, and an ending inventory of $40,000. Compute the average inventory.

SOLUTION: $35,000

56.
A merchant had a beginning inventory with a retail value of $30,000. During the year the merchant purchased goods with a retail value of $150,000. At year-end he merchant had inventory with a retail value of $40,000. Sales for the year were $140,000. Compute the inventory turnover at retail.

SOLUTION: 4.0

57.
Lambert's Auto Mart had a beginning inventory of $1,400,800, purchases of $3,078,900, and an ending inventory of $1,600,500. Compute the average inventory.

SOLUTION: $1,500,650

58.
Lambert's Auto Mart had a beginning inventory of $1,400,800, purchases of $3,001,300 and an ending inventory of $1,600,500. Compute the inventory turnover.

SOLUTION: 2

59.

Keating Auto Mart had a beginning inventory of $2,000,000 at retail. Keating had quarterly inventories of $7,000,000 at the end of Q1, $7,000,000 at the end of Q2, $5,000,000 at the end of Q3, and $6,000,000 at year-end. Compute the average inventory.

SOLUTION: $5,400,000

60.

If the year-end inventory for Keating in problem 59 had been the same as the beginning of the year inventory, what would have been the average inventory for the year?

SOLUTION: $4,600,000

Chapter 18 DEPRECIATION

PROBLEMS

LEARNING OBJECTIVE 1

1.
Michigan Manufacturing uses the straight-line method of depreciation. A large fork-lift was purchased for $62,500. It has an estimated life of five years and an estimated scrap value of $2,500. Compute the annual depreciation amount.

SOLUTION: $12,000

2.
From the data in problem 1, compute the book value at the end of the first year.

SOLUTION: $50,500

3.
Voice Genesis Inc. uses the units of production method of depreciation. Electronic equipment costing $52,500 has 60,000 estimated hours of operation and an estimated scrap value of $2,500. It operated 18,000 hours in the first year. Compute the depreciation expense for the first year.

SOLUTION: $15,000

4.
From the data in problem 3, compute the book value at the end of the first year.

SOLUTION: $37,500

5.
Northgate Coal Mining uses the units of production method of depreciation. A piece of heavy equipment costing $368,000 produces an estimated 1,740,000 units in its life and has an estimated scrap value of $20,000. It produced 316,000 units this year. Compute the depreciation for the first year.

SOLUTION: $63,200

6.
From the data in problem 5, compute the book value at the end of the first year.

SOLUTION: $304,800

7.
Apex Electronics uses the straight-line method of depreciation. Electronic equipment costing $112,500 has an estimated life of ten years and an estimated scrap value of $8,100. Compute the annual depreciation amount.

SOLUTION: $10,440

8.

From the data in problem 7, compute the book value at the end of the third year.

SOLUTION: $81,180

9.

Mendocino Gas and Storage uses the units of production method of depreciation. A piece of equipment costing $112,500 has 360,000 estimated hours of operation and an estimated scrap value of $8,100. It operated 56,000 hours in the first year. Compute the depreciation expense that will be shown for the first year.

SOLUTION: $16,240

10.

From the data in problem 9, compute the book value at the end of the first year.

SOLUTION: $96,260

11.

Miracle Manufacturing uses the units of production method of depreciation. A piece of heavy equipment costing $112,500 produces an estimated 2,088,000 units in its life and has an estimated scrap value of $8,100. It produced 288,000 units this year. Compute the depreciation for the first year.

SOLUTION: $14,400

12.

From the data in problem 11, compute the book value at the end of the first year.

SOLUTION: $98,100

13.

Rolex Watch Company uses the straight-line method of depreciation. A new lathe costing $22,500 has an estimated life of five years and an estimated scrap value of $2,500. Compute the annual depreciation amount.

SOLUTION: $4,000

14.

From the data in problem 13, compute the book value of the equipment at the end of the third year.

SOLUTION: $10,500

15.

Perry Processing Inc. uses the units of production method of depreciation. A packaging machine costing $22,500 has 40,000 estimated hours of operation and an estimated scrap value of $2,500. It operated 4,200 hours in the first year. Compute the depreciation expense for the first year.

SOLUTION: $2,100

16.
From the data in problem 15, compute the book value at the end of the first year.

SOLUTION: $20,400

17.
Mason Towel uses the units of production method of depreciation. A new knitting machine was purchased for $22,500. It will produce an estimated 800,000 units in its life and has an estimated scrap value of $2,500. It produced 150,000 units in the first year. Compute the depreciation for the year.

SOLUTION: $3,750

18.
From the data in problem 15, compute the book value at the end of the first year.

SOLUTION: $18,750

19.
Cole Camping Company uses the double-declining balance method of depreciation. A piece of equipment costing $37,500 has an estimated life of five years and an estimated scrap value of $2,700. Compute the amount of depreciation taken in the second year.

SOLUTION: $9,000

20.
From the data in problem 19, compute the book value at the end of the second year.

SOLUTION: $13,500

21.
Bellinger Broadcasting uses the double-declining balance method of depreciation. A transmitter costing $112,500 has an estimated life of ten years and an estimated scrap value of $8,100. Compute the amount of depreciation taken in the third year.

SOLUTION: $14,400

22.
From the data in problem 21, compute the book value at the end of the third year.

SOLUTION: $57,600

23.
Zenix Electronic uses the double-declining balance method of depreciation. An oscilloscope costing $28,500 has an estimated life of five years and an estimated scrap value of $2,500. Compute the amount of depreciation taken in the second year.

SOLUTION: $6,840

24.

From the data in problem 23, compute the book value at the end of the second year.

SOLUTION: $10,260

25.

Brook Stone and Paving purchased a new truck for $68,500. The truck has an estimated life of 4 years and an estimated scrap value of $4,500. BSP uses the double-declining balance method of depreciation. Compute the depreciation for year 3.

SOLUTION: $8,562.5

26.

From the data in problem 25, compute the book value at the end of the third year.

SOLUTION: $8,562.5

27.

Young Manufacturing uses the sum-of-the-years-digits method of depreciation. A piece of heavy equipment costing $37,500 has an estimated life of five years and an estimated scrap value of $2,700. Compute the amount of depreciation taken in the third year.

SOLUTION: $6,960

28.

Young Manufacturing uses the sum-of-the-years-digits method of depreciation. A piece of heavy equipment costing $37,500 has an estimated life of five years and an estimated scrap value of $2,700. Compute the book value at the end of the second year.

SOLUTION: $16,620

29.

Warner Brothers Studios uses the sum-of-the-years-digits method of depreciation. A camera unit costing $112,500 has an estimated life of ten years and an estimated scrap value of $8,100. Compute the amount of depreciation taken in the seventh year.

SOLUTION: $7,592.73

30.

From the data in problem 27, compute the book value at the end of the third year. (Round depreciation amounts to the nearest dollar.)

SOLUTION: $61,249

31.

The bookkeeper for the Martel Company is computing depreciation for income tax purposes, using the figures from the MACRS tables supplied by the IRS. The equipment being depreciated had a cost of $16,000 and falls under the class of equipment to be depreciated at a rate of 25% the first year and 21.43% the second. The equipment was purchased and put into use during the first quarter. Compute the amount of depreciation expense for the first year.

SOLUTION: $4,000

32.

The bookkeeper for the Martel Company is computing depreciation for income tax purposes, using the figures from the MACRS tables supplied by the IRS. The equipment being depreciated had a cost of $16,000 and falls under the class of equipment to be depreciated at a rate of 20.00% the first year and 32.00% the second. The equipment was purchased and put into use during the first quarter. Compute the total amount of depreciation expense through the end of the second year. (Round to the nearest dollar.)

SOLUTION: $8,320

33.

The bookkeeper for the Reilly Construction Company is computing depreciation for income tax purposes, using the figures from the MACRS tables supplied by the IRS. Reilly Construction Company purchased a truck in February 2005, for $40,000. In March, 2005, a second truck was purchased at $48,000. Assume that allowable depreciation for each truck is 20.00% the first year and 32.00% the second. Compute the total allowable cost recovery on the two trucks for the year 2005.

SOLUTION: $17,600 ($40,000 + $48,000) × .2.

34.

The bookkeeper for the Reilly Construction Company is computing depreciation for income tax purposes, using the figures from the MACRS tables supplied by the IRS. Reilly Construction Company purchased a truck in February 2005 for $40,000. In March 2005 a second truck was purchased at $48,800. Assume that allowable depreciation for each truck is 20.00% the first year and 32.00% the second. Compute the total allowable cost recovery on the two trucks for the year 2006.

SOLUTION: $28,416   ($40,000 + $48,800) × .32

LEARNING OBJECTIVE 5

35.

A dental office purchased equipment costing $72,000 and put it into use June 1. The equipment is expected to have a useful life of 10 years and an estimated resale value of $4,800. Using the straight-line method of depreciation, compute the depreciation expense for June 1 through December 31 of the first tax year and all 12 months of each of the second and third years.

SOLUTION: $17,360

36.

A dental office purchased equipment costing $72,000 and put it into use June 1. The equipment is expected to have a useful life of 10 years and an estimated resale value of $4,800. Using the straight-line method of depreciation, compute the book value at end of the third year.

SOLUTION: $54,640

37.

A day-care center purchased equipment costing $36,000 and put it into use June 1. The equipment is expected to have a useful life of 10 years and an estimated resale value of $2,400. Using the double-declining-balance method of depreciation, compute the depreciation expense for June 1 through December 31 of the first tax year and all 12 months of each of the second and third years.

SOLUTION: $15,648

38.

A day-care center purchased equipment costing $36,000 and put it into use June 1. The equipment is expected to have a useful life of 10 years and an estimated resale value of $2,400. Using the double-declining-balance method of depreciation, compute the book value at the end of the third year.

SOLUTION: $20,352

39.

A retail store purchased equipment costing $36,000 and put it into use June 1. The equipment is expected to have a useful life of 10 years and an estimated resale value of $2,400. Using the sum-of-the-years-digits method of depreciation, compute the depreciation expense for June 1 through December 31 of the first year and all 12 months of the second year. (Round depreciation to the nearest dollar.)

SOLUTION: $9,316

40.

A retail store purchased equipment costing $36,000 and put it into use June 1. The equipment is expected to have a useful life of 10 years and an estimated resale value of $2,400. Using the sum-of-the-years-digits method of depreciation, compute the book value at the end of the second year. (Round depreciation to the nearest dollar.)

SOLUTION: $26,684

41.

An insurance office purchased office furniture costing $9,600 and put it into use April 1. The furniture is expected to have a useful life of 10 years and an estimated resale value of $600. Using the straight-line method of depreciation, compute the depreciation expense for April 1 through December 31 of the first year and all of each of the second year.

SOLUTION: $675 and $900 ($9,600 – $600= 9,000/10 = $900)

42.

An insurance office purchased office furniture costing $9,000 and put it into use April 1. The furniture is expected to have a useful life of 10 years and an estimated resale value of $600. Using the straight-line method of depreciation, compute the book value at the end of the third year.

SOLUTION: $6,690

43.

A real estate office purchased office furniture costing $9,000 and put it into use April 1. The furniture is expected to have a useful life of 10 years and an estimated resale value of $600. Using the double-declining-balance method of depreciation, compute the depreciation expense for April 1 through December 31 of the first tax year and all 12 months of each of the second and third years.

SOLUTION: $4,104

44.

A real estate office purchased office furniture costing $9,000 and put it into use April 1. The furniture is expected to have a useful life of 10 years and an estimated resale value of $600. Using the double-declining-balance method of depreciation, compute the book value at the end of the third year.

SOLUTION: $4,896

45.

An accounting firm purchased office furniture costing $9,000 and put it into use April 1. The furniture is expected to have a useful life of 10 years and an estimated resale value of $600. Using the sum-of-the-years-digits method of depreciation, compute the depreciation expense for April 1 through December 31 of the first tax year and all 12 months of the second year. (Round depreciation for each year to the nearest dollar.)

SOLUTION: $2,558

46.

An accounting firm purchased office furniture costing $9,000 and put it into use April 1. The furniture is expected to have a useful life of 10 years and an estimated resale value of $600. Using the sum-of-the-years-digits method of depreciation, compute the book value at the end of the second year. (Round depreciation for each year to the nearest dollar.)

SOLUTION: $6,442

47.

A legal office purchased office furniture costing $36,000 and put it into use May 1. The furniture is expected to have a useful life of 10 years and an estimated resale value of $2,400. Using the straight-line method of depreciation, compute the depreciation expense for May 1 through December 31 of the first tax year and all 12 months of each of the second and third years.

SOLUTION: $8,960

48.

A legal office purchased office furniture costing $36,000 and put it into use May 1. The furniture is expected to have a useful life of 10 years and an estimated resale value of $2,400. Using the straight-line method of depreciation, compute the book value at the end of the third year.

SOLUTION: $27,040

49.

A medical office purchased office furniture costing $36,000 and put it into use May 1. The furniture is expected to have a useful life of 10 years and an estimated resale value of $2,400. Using the double-declining-balance method of depreciation, compute the depreciation expense for May 1 through December 31 of the first tax year and all 12 months of each of the second and third years.

SOLUTION: $16,032

50.

A medical office purchased office furniture costing $36,000 and put it into use May 1. The furniture is expected to have a useful life of 10 years and an estimated resale value of $2,400. Using the double-declining-balance method of depreciation, compute the book value at the end of the third year.

SOLUTION: $19,968

51.

A publisher purchased office furniture costing $36,000 and put it into use May 1. The furniture is expected to have a useful life of 10 years and an estimated resale value of $2,400. Using the sum-of-the-years-digits method of depreciation, compute the depreciation expense for May 1 through December 31 of the first year and all 12 months of the second year. (Round depreciation for each year to nearest dollar.)

SOLUTION: $9,775

52.

A publisher purchased office furniture costing $36,000 and put it into use May 1. The furniture is expected to have a useful life of 10 years and an estimated resale value of $2,400. Using the sum-of-the-years-digits method of depreciation, compute the book value at the end of the second year.

SOLUTION: $26,225

53 – 60.

The California Crane Company purchased eight new cranes of different sizes and qualities. CCC uses the double-declining balance method of calculating depreciation. The data for each crane is given below. From this data compute the 1st year deprecation for each crane and the value of each crane at the end of its 1st year of use.

Cost	Est.Life	Scrap Value	Depreciation Year 1	Value at End of 1st Year
$220,000	10 years	$22,000	_____	_____
$316,000	8 years	$31,600	_____	_____
$ 98,400	10 years	$9,840	_____	_____
$420,000	10 years	$42,000	_____	_____
$108,800	6 years	$10,800	_____	_____
$784,500	8 years	$78,450	_____	_____
$780,000	8 years	$78,000	_____	_____
$520,000	4 years	$52,000	_____	_____

SOLUTION:

	Depreciation Year 1	Value at end of 1st Year
53.	$88,000	$132,000
54.	$79,000	$237,000
55.	$19,680	$78,720
56.	$84,000	$336,000
57.	$36,227	$72,533
58.	$196,125	$588,375
59.	$195,000	585,000
60.	$260,000	$260,000

Chapter 19 FINANCIAL STATEMENTS

PROBLEMS

LEARNING OBJECTIVES 1, 2

1.
Blackstone Buildings, Inc. had current liabilities of $450,600; cash of $310,360; accounts receivable of $165,242; and ending inventory of $210,450. Compute its total current assets.

SOLUTION: $686,052

2.
Blackstone Buildings, Inc. had net sales of $432,600, cost of goods sold of $308,100, and payroll expense of $166,000. Compute the gross profit on sales.

SOLUTION: $124,500

3.
Blackstone Buildings, Inc. had net sales of $432,600; cost of goods sold of $308,100; total assets of $843,750; and total liabilities of $768,290. Compute the owners' equity.

SOLUTION: $75,460

4.
Zucker Global Products, Inc. had current liabilities of $453,190; cash of $220,250; accounts receivable of $100,350; and ending inventory of $225,670. Compute the total current assets.

SOLUTION: $546,270

5.
Zucker Global Products, Inc. had net sales of $299,999 and cost of goods sold of $237,545. Compute the gross profit on sales.

SOLUTION: $62,454

6.
Zucker Global Products, Inc. had net sales of $299,999; net income of $20,600; total assets of $795,180; and total liabilities of $602,100. Compute the rate of return on investment. Express your answer as a percentage. (Compute answer accurate to two decimal places.)

SOLUTION: 10.67%

7.
The comparative income statement of Lawn Products, Inc. showed sales of $325,000 in 2004 and $494,450 in 2005. Compute the dollar amount of net change in sales.

SOLUTION: $169,450

8.

The comparative income statement of Lawn Products, Inc. showed sales of $325,000 in 2004 and $494,450 in 2005. Compute the percentage of net change in sales. (Round answer to one decimal place.)

SOLUTION: 52.1%

9.

The comparative income statement of Lawn Products, Inc. showed net income of $24,500 in 2004 and $27,300 in 2005. Compute the dollar amount of net change in net income.

SOLUTION: $2,800

10.

The comparative income statement of Lawn Products, Inc. showed net income of $24,500 in 2004 and $27,300 in 2005. Compute the percentage of net change in net income. (Round answer to one decimal place.)

SOLUTION: 11.4%

11.

The balance sheet of Lawn Products, Inc. showed accounts receivable of $84,750 in 2004 and $134,750 in 2005. Compute the percentage of increase in accounts receivable. (Round answer to one decimal place.)

SOLUTION: 59.0%

12.

The balance sheet of Lawn Products, Inc. showed owners' equity of $180,000 in 2001 and $200,000 in 2002. Compute the percentage of increase in owners' equity. (Round answer to one decimal place.)

SOLUTION: 11.1%

13.

Selected figures from the Balance Sheet and Income Statement of Multiple Enterprises, Inc. are shown below. Compute the working capital ratio. (Compute answer accurate to two decimal places.)

From the Balance Sheet		From the Income Statement	
Cash	$ 514,800	Net Sales	$1,900,000
Accounts Receivable	$ 324,000	Cost of Goods Sold..	$ 796,000
Merchandise Inventory:		Net Income	$ 240,000
End of this year	$ 498,000		
End of last year	$ 432,000		
Total Current Assets	$1,336,800		
Total Current Liabilities.	$ 269,200		
Total Stockholders' Equity	$1,292,800		

SOLUTION: 4.97:1

14.

Selected figures from the Balance Sheet and Income Statement of Multiple Enterprises, Inc. are shown below. Compute the acid test ratio. (Compute answer accurate to two decimal places.)

From the Balance Sheet		From the Income Statement	
Cash	$ 514,800	Net Sales..................	$1,900,000
Accounts Receivable............	$ 324,000	Cost of Goods Sold..	$ 796,000
Merchandise Inventory:		Net Income	$ 240,000
End of this year	$ 498,000		
End of last year........	$ 432,000		
Total Current Assets.............	$1,336,800		
Total Current Liabilities.......	$ 269,200		
Total Stockholders' Equity...	$1,292,800		

SOLUTION: 3.12:1

15.

Selected figures from the Balance Sheet and Income Statement of Multiple Enterprises, Inc. are shown below. Compute the inventory turnover rate. (Compute answer accurate to two decimal places.)

From the Balance Sheet		From the Income Statement	
Cash	$ 514,800	Net Sales..................	$1,900,000
Accounts Receivable............	$ 324,000	Cost of Goods Sold..	$ 796,000
Merchandise Inventory:		Net Income	$ 240,000
End of this year	$ 498,000		
End of last year........	$ 432,000		
Total Current Assets.............	$1,336,800		
Total Current Liabilities.......	$ 269,200		
Total Stockholders' Equity...	$1,292,800		

SOLUTION: 1.71

16.

Selected figures from the Balance Sheet and Income Statement of Multiple Enterprises, Inc. are shown below. Compute the ratio of accounts receivable to net sales. (Compute answer accurate to two decimal places.)

From the Balance Sheet		From the Income Statement	
Cash	$ 514,800	Net Sales..................	$1,900,000
Accounts Receivable............	$ 324,000	Cost of Goods Sold..	$ 796,000
Merchandise Inventory:		Net Income	$ 240,000
End of this year	$ 498,000		
End of last year........	$ 432,000		
Total Current Assets.............	$1,336,800		
Total Current Liabilities.......	$ 269,200		
Total Stockholders' Equity...	$1,292,800		

SOLUTION: 0.17:1

17.
Selected figures from the Balance Sheet and Income Statement of Multiple Enterprises, Inc. are shown below. Compute the relationship of net income to net sales. Express your answer as a percentage. (Compute answer accurate to two decimal places.)

From the Balance Sheet	From the Income Statement
Cash .... $ 514,800	Net Sales................. $1,900,000
Accounts Receivable............ $ 324,000	Cost of Goods Sold.. $ 796,000
Merchandise Inventory:	Net Income .... $ 240,000
End of this year .... $ 498,000	
End of last year........ $ 432,000	
Total Current Assets............. $1,336,800	
Total Current Liabilities....... $ 269,200	
Total Stockholders' Equity... $1,292,800	

SOLUTION: 12.63%

18.
Selected figures from the Balance Sheet and Income Statement of Multiple Enterprises, Inc. are shown below. Compute the rate of return on investment. Express your answer as a percentage. (Compute answer accurate to two decimal places.)

From the Balance Sheet	From the Income Statement
Cash .... $ 514,800	Net Sales................. $1,900,000
Accounts Receivable............ $ 324,000	Cost of Goods Sold.. $ 796,000
Merchandise Inventory:	Net Income .... $ 240,000
End of this year .... $ 498,000	
End of last year........ $ 432,000	
Total Current Assets............. $1,336,800	
Total Current Liabilities....... $ 269,200	
Total Stockholders' Equity... $1,292,800	

SOLUTION: 18.56%

19.
Selected figures from the Balance Sheet and Income Statement of Turner's Toys, Inc. are shown below. Compute the working capital ratio. (Compute answer accurate to two decimal places.)

From the Balance Sheet	From the Income Statement
Cash .... $ 210,734	Net Sales.................... $244,750
Accounts Receivable............ $ 138,126	Cost of Goods Sold..... $190,000
Merchandise Inventory:	Net Income .... $ 26,406
End of this year .... $ 184,500	
End of last year........ $ 178,300	
Total Current Assets............. $ 533,360	
Total Current Liabilities....... $ 324,152	
Total Stockholders' Equity... $ 149,208	

SOLUTION: 1.65:1

20.

Selected figures from the Balance Sheet and Income Statement of Turner's Toys, Inc. are shown below.
Compute the acid test ratio. (Compute answer accurate to two decimal places.)

From the Balance Sheet		From the Income Statement	
Cash	$ 210,734	Net Sales	$244,750
Accounts Receivable	$ 138,126	Cost of Goods Sold	$190,000
Merchandise Inventory:		Net Income	$ 26,406
End of this year	$ 184,500		
End of last year	$ 178,300		
Total Current Assets	$ 533,360		
Total Current Liabilities	$ 324,152		
Total Stockholders' Equity	$ 149,200		

SOLUTION: 1.08:1

21.

Selected figures from the Balance Sheet and Income Statement of Turner's Toys, Inc. are shown below.
Compute the inventory turnover rate. (Compute answer accurate to two decimal places.)

From the Balance Sheet		From the Income Statement	
Cash	$ 210,734	Net Sales	$244,750
Accounts Receivable	$ 138,126	Cost of Goods Sold	$190,000
Merchandise Inventory:		Net Income	$ 26,406
End of this year	$ 184,500		
End of last year	$ 178,300		
Total Current Assets	$ 533,360		
Total Current Liabilities	$ 324,152		
Total Stockholders' Equity	$ 149,200		

SOLUTION: 1.05

22.

Selected figures from the Balance Sheet and Income Statement of Turner's Toys, Inc. are shown below.
Compute the ratio of accounts receivable to net sales. (Compute answer accurate to two decimal places.)

From the Balance Sheet		From the Income Statement	
Cash	$ 210,734	Net Sales	$244,750
Accounts Receivable	$ 138,126	Cost of Goods Sold	$190,000
Merchandise Inventory:		Net Income	$ 26,406
End of this year	$ 184,500		
End of last year	$ 178,300		
Total Current Assets	$ 533,360		
Total Current Liabilities	$ 324,152		
Total Stockholders' Equity	$ 149,200		

SOLUTION: 0.56:1

23.
Selected figures from the Balance Sheet and Income Statement of Turner's Toys, Inc. are shown below. Compute the relationship of net income to net sales. Express your answer as a percentage. (Compute answer accurate to two decimal places.)

From the Balance Sheet		From the Income Statement	
Cash	$ 210,734	Net Sales	$244,750
Accounts Receivable	$ 138,126	Cost of Goods Sold	$190,000
Merchandise Inventory:		Net Income	$ 26,406
End of this year	$ 184,500		
End of last year	$ 178,300		
Total Current Assets	$ 533,360		
Total Current Liabilities	$ 324,152		
Total Stockholders' Equity	$ 149,200		

SOLUTION: 10.79%

24.
Selected figures from the Balance Sheet and Income Statement of Turner's Toys, Inc. are shown below. Compute the rate of return on investment. Express your answer as a percentage. (Compute answer accurate to two decimal places.)

From the Balance Sheet		From the Income Statement	
Cash	$ 210,734	Net Sales	$244,750
Accounts Receivable	$ 138,126	Cost of Goods Sold	$190,000
Merchandise Inventory:		Net Income	$ 26,406
End of this year	$ 184,500		
End of last year	$ 178,300		
Total Current Assets	$ 533,360		
Total Current Liabilities	$ 324,152		
Total Stockholders' Equity	$ 149,200		

SOLUTION: 17.72%

25.
The Moose Hardware Company had total current assets of $379,090 and total current liabilities of $63,000. Compute the working capital ratio. (Compute answer accurate to two decimal places.)

SOLUTION: 6.02:1

26.
The Moose Hardware Company had cash of $160,000; current receivables of $136,000; and current liabilities of $63,000. Compute the acid test ratio.

SOLUTION: 4.70:1

27.
The Moose Hardware Company had ending inventory of $143,600 this year, ending inventory of $131,000 last year, and $260,000 as cost of goods sold. Compute the inventory turnover rate.

SOLUTION: 1.89

28.

The Moose Hardware Company had accounts receivable of $136,000 and net sales of $390,000. Compute the ratio of accounts receivable to net sales. (Compute answer accurate to two decimal places.)

SOLUTION: 0.35:1

29.

The Moose Hardware Company had net income of $60,000 and net sales of $390,000. Compute the relationship of net income to net sales. Express your answer as a percentage. (Compute answer accurate to two decimal places)

SOLUTION: 15.38%

30.

The Moose Hardware Company had net income of $60,000 and owners' equity of $216,000. Compute the rate of return on investment. Express your answer as a percentage. (Compute answer accurate to two decimal places.)

SOLUTION: 27.78%

31.

Thompson Laser Company had current assets of $402,090 and current liabilities of $104,000. Compute the working capital ratio. (Compute answer accurate to two decimal places.)

SOLUTION: 3.87:1

32.

Thompson Laser Company had cash of $160,000; accounts receivable of $136,000; and current liabilities of $104,000. Compute the acid test ratio.

SOLUTION: 2.85:1

33.

Thompson Laser Company had ending inventory of $163,600 this year, ending inventory of $134,000 last year, and $262,000 as cost of goods sold. Compute the inventory turnover rate.

SOLUTION: 1.76

34.

Forman and Brasso Furniture Company had total current assets of $455,000 and total current liabilities of $295,400. Compute the working capital ratio. (Compute answer accurate to two decimal places.)

SOLUTION: 1.54:1

35.

Forman and Brasso Furniture Company had cash of $182,400; accounts receivable of $126,200; and current liabilities of $295,400. Compute the acid test ratio.

SOLUTION: 1.04:1

36.
Forman and Brasso Furniture Company had ending inventory of $146,400 this year, ending inventory of $152,200 last year, and $194,400 as cost of goods sold. Compute the inventory turnover rate.

SOLUTION: 1.30

37.
Forman and Brasso Furniture Company had accounts receivable of $126,200 and net sales of $248,800. Compute the ratio of accounts receivable to net sales. (Compute answer accurate to two decimal places.)

SOLUTION: 0.51:1

38.
Forman and Brasso Furniture Company had net income of $34,000 and net sales of $248,800. Compute the relationship of net income to net sales. Express your answer as a percentage. (Compute answer accurate to two decimal places.)

SOLUTION: 13.67%

39.
Forman and Brasso Furniture Company had net income of $34,000 and owners' equity of $130,500. Compute the rate of return on investment. Express your answer as a percentage. (Compute answer accurate to two decimal places.)

SOLUTION: 26.05%

40.
Quality Construction, Inc. had current liabilities of $450,600; cash of $280,368; accounts receivable of $135,242; and ending inventory of $210,450. Compute the working capital ratio. (Compute answer accurate to two decimal places.)

SOLUTION: 1.39:1

41.
Quality Construction, Inc. had current liabilities of $450,600; cash of $280,368; accounts receivable of $135,242; and ending inventory of $210,450. Compute the acid test ratio. (Compute answer accurate to two decimal places.)

SOLUTION: 0.92:1

42.
Quality Construction, Inc. had ending inventory this year of $210,450; ending inventory last year of $196,350; and cost of goods sold of $308,100. Compute the average inventory.

SOLUTION: $203,400

43.

Quality Construction, Inc. had ending inventory this year of $210,450; ending inventory last year of $196,350; and cost of goods sold of $308,100. Compute the inventory turnover rate. (Compute answer accurate to two decimal places.)

SOLUTION: 1.51

44.

Quality Construction, Inc. had accounts receivable of $135,242 and net sales of $408,140. Compute the ratio of accounts receivable to net sales. (Compute answer accurate to two decimal places.)

SOLUTION: 0.33:1

45.

Quality Construction, Inc. had net sales of $408,140 and net income of $47,000. Compute the relationship of net income to net sales. Express your answer as a percentage. (Compute answer accurate to two decimal places.)

SOLUTION: 11.52%

46.

Quality Construction, Inc. had net sales of $408,140, net income of $47,000; total assets of $943,750; and total liabilities of $768,290. Compute the rate of return on investment. Express your answer as a percentage. (Compute answer accurate to two decimal places.)

SOLUTION: 26.79%

47.

Warm Winters Linen, Inc. had current liabilities of $500,000; cash of $300,000; accounts receivable of $150,000; and ending inventory of $250,000. Compute the working capital ratio. (Compute answer accurate to two decimal places.)

SOLUTION: 1.4:1

48.

Warm Winters Linen, Inc. had current liabilities of $500,000; cash of $300,000; accounts receivable of $150,000; and ending inventory of $250,000. Compute the acid test ratio. (Compute answer accurate to two decimal places.)

SOLUTION: 0.9:1

49.

Warm Winters Linen, Inc. had ending inventory this year of $250,000; ending inventory last year of $200,000; and cost of goods sold of $240,000. Compute the average inventory.

SOLUTION: $225,000

50.
Warm Winters Linen, Inc. had ending inventory this year of $250,000; ending inventory last year of $200,000; and cost of goods sold of $240,000. Compute the inventory turnover. (Compute answer accurate to two decimal places.)

SOLUTION: 1.07

51.
Warm Winters Linen, Inc. had accounts receivable of $150,000 and net sales of $350,000. Compute the ratio of accounts receivable to net sales. (Compute answer accurate to two decimal places.)

SOLUTION: 0.43:1

52.
Warm Winters Linen, Inc. had net sales of $350,000 and net income of $30,000. Compute the relationship of net income to net sales. Express your answer as a percentage. (Compute answer accurate to two decimal places.)

SOLUTION: 8.57%

53. – 56.
The Income Statement for Mentura Products for the Month and Three-Month Period Ended March 31, 20xx is shown below. Show the Amount and Percent of Differences from the Budgeted Amounts for the Month and the Three-Month Periods.

Mentura Products
Income Statement for the Month and the
Three-Month Period Ended June 30, 20xx.

| | Month of March | | | | Three Months | | | |
	Budget	Actual	Diff	Amount Diff	% Budget	Actual	Amount Diff	% Diff
Sales								
Sales	$26,000	$24,000	____	____	$84,000	$90,000	____	____
Returns	3,000	2,000	____	____	7,000	6,000	____	____
Net Sales	$23,000	$22,000	____	____	$77,000	$84,000	____	____
Expenses								
Production	4,000	5,000	____	____	14,000	12,000	____	____
General	8,000	9,000	____	____	26,000	24,000	____	____
Other	3,000	4,000	____	____	10,000	11,000	____	____
Total	$15,000	$18,000	____	____	$50,000	$47,000	____	____
Net Income	$ 8,000	$ 4,000	____	____	$27,000	$37,000	____	____

SOLUTION:

Month Amount Diff	Month Percent Diff	Three-Month Amount Diff	Three-Month Percent Diff
($2,000)	(0.08%)	$6,000	0.07%
($1,000)	(0.33%)	($1,000)	(0.14%)
($1,000)	(0.04%)	$7,000	0.09%
$1,000	0.25%	($2,000)	(0.14%)
$1,000	0.13%	($2,000)	(0.08%)
$1,000	0.33%	$1,000	0.10%
$3,000	0.20%	($3,000)	(0.06%)
($4,000)	(0.50%)	$10,000	0.37%

Chapter 20 INTERNATIONAL BUSINESS

PROBLEMS

LEARNING OBJECTIVE 1

1.
If the Australian dollar exchange rate in U.S. Dollars is 0.7253, compute the amount of U.S. currency required to buy 200 Australian Dollars.

SOLUTION: $145.06 U.S. Dollars

2.
If the Australian Dollar exchange rate in U.S. Dollars is 0.7253, compute the amount in Australian Dollars needed to purchase $200 in U.S. currency.

SOLUTION: $275.75 (Australian Dollars)

3.
An American purchased a shirt in Melbourne, Australia for 42 Australian Dollars. The exchange rate was 0.7259 Australian Dollars for one U.S. Dollar. How much did he have to pay in U.S. Dollars?

SOLUTION: $57.86

4.
An American purchased a pair of shoes in Sydney, Australia.   The price was 120 Australian Dollars plus a 6% tax. How much did he have to pay in American Dollars when the exchange rate for Australian Dollars was 0.7286 per U.S. Dollar?

SOLUTION: $174.58

5.
An Australian firm exported manufactured electronic components to the United States valued at 54,800 Australian Dollars.   The exchange rate for Australian Dollars was 0.7654 per U.S. Dollar. What was the amount of U.S. Dollars the exporter received?

SOLUTION: $71,596.55

6.
If the exchange rate for the Australian Dollar is 0.8425 per U.S. Dollar, compute the value in U.S. Dollars of 5,000 Australian Dollars.

SOLUTION: $4,212.50 in U.S. Dollars

7.
If the Mexican Peso is valued at 11.0255 per U.S. Dollar, compute the amount of U.S. currency necessary to buy 100 Mexican Pesos.

SOLUTION: $9.07 U.S. Dollars

8.

If the Mexican Peso is listed under Currency Exchange Rates as 0.08716 in U.S. Dollars, compute the amount in Mexican Pesos one could purchase for $3,100 in U.S. currency.

SOLUTION: 35,566.77

9.

If the Mexican Peso is valued at 11.0255 per U.S. Dollar, compute the value in U.S. currency of 5,000 Mexican Pesos.

SOLUTION: $453.49 in U.S. Dollars

10.

A tourist from the United States wants to purchase a painting in Mexico City. The price of the painting was 5,927 pesos plus a shipping charge of 387 pesos to send the painting to the tourist's home, how many U.S. dollars would the tourist have to pay when the exchange rate was 11.0255 pesos per U.S. Dollar?

SOLUTION: $572.67

11.

If the Brazilian Real is valued at 2.502 per U.S. Dollar, compute the amount of Brazilian Reals one could purchase for $100 in U.S. currency.

SOLUTION: 250.2 Brazilian Reals

12.

If the Brazilian Real is valued at 2.502 per U.S. Dollar, compute the value in U.S. currency of 5,000 Brazilian Reals.

SOLUTION: $1,998.40 in U.S. Dollars

13.

If the Brazilian Real is valued at 0.3997 U.S. Dollars, how much in U.S. Dollars would be required to satisfy an invoice for produce shipped from Brazil with a price of 24,000 Brazilian Reals?

SOLUTION: $9,592.80

14.

If the Brazilian Real is valued at 0.3997 to the U.S. Dollar, compute the amount of Brazilian Reals needed to pay for purchase of a television set costing $1,222.22.

SOLUTION: 3,057.84

15.

If the Danish Krone is listed under Foreign Currency in Dollars as 0.1731, compute the amount of U.S. Dollars necessary to buy 100 Danish Krone.

SOLUTION: $17.31

16.
If the Danish Krone has a current value of 5.7757 per U.S. Dollar, compute the amount that could be purchased for $350.

SOLUTION: 2,021.49

17.
If the Danish Krone is valued at 5.7757 per U.S. Dollar how many dollars would be required to purchase a Danish rail ticket costing 748.50 Danish Krone?

SOLUTION: $129.59

18.
An American import company purchased 700 cans of Danish cookies at a price of 8.20 krone per can. If the Danish Krone were valued at 5.7757 on the day of settlement, how much did the company pay in U.S. Dollars?

SOLUTION: $993.82

19.
If an American import company purchased three gross of Danish hams at a price of 14.40 krone per ham, and the Danish Krone were valued at 5.7757 on the day of settlement, how much did the company pay in U.S. Dollars?

SOLUTION: $1,077.06

20.
If the U.S. Dollar is valued at 0.1731 per Danish Krone, how many dollars would a company pay to purchase merchandise from Denmark costing 70,148 Danish Krone?

SOLUTION: $12,142.61

21.
If the Euro is valued at $1.2936 in U.S. dollars, compute the amount of U.S. Dollars necessary to buy 100 Euro.

SOLUTION: $129.36

22.
If the Euro is valued at $1.2936 in U.S. dollars, compute the amount in Euro one could purchase for $100 in U.S. currency. (Round to two places.)

SOLUTION: 77.30

23.
If the Euro is valued at $1.2936 in U.S. dollars, compute the value in U.S. currency of 5,000 Euro.

SOLUTION: $6,468.00

24.

If the U.S. Dollar is valued at $1.8910 per British Pound, compute the amount of U.S. currency necessary to buy 200 British Pounds.

SOLUTION: $378.20

25.

If the British Pound is valued at 0.5288 per U.S. Dollar, compute the amount of pounds one could purchase for $500 in U.S. currency. (Round answer to two decimal places.)

SOLUTION: 264.40

26.

If the British Pound is valued at 0.5288 per U.S. Dollar compute the value in U.S. currency of 7,000 pounds.

SOLUTION: $13,237.52

27.

If the British Pound is valued at 0.5288 per U.S. Dollar, how much would an American import company pay in U.S. Dollars for merchandise valued at 789,400 pounds?

SOLUTION: $1,492, 813.90

28.

If the exchange rate is 0.5288 British Pound for 1 U.S. Dollar, compute the amount of pounds required to pay an American export company for $48,900 in merchandise. (Round answer to the nearest whole number.)

SOLUTION: 25,858

29.

If the exchange rate is 0.5288 British Pound for 1 U.S. Dollar, how many pounds would it cost a British merchant to pay an American export company for merchandise valued at $200,000?   (Round to nearest whole number.)

SOLUTION: 105,760

30.

If the U.S. Dollar is worth 105.16 yen, compute the amount of U.S. Dollars necessary to buy 50,000 Japanese yen.

SOLUTION: $475.47

31.

If the U.S. Dollar is worth 105.16 yen, compute the amount a U.S. company would have to pay to buy merchandise valued at 3,900,000 Japanese yen.

SOLUTION: $37,086.34

32.

If the Japanese yen is worth 0.009545 U.S. Dollar, what is the dollar value of 100,000 yen?

SOLUTION: $954.50

33.

If the Japanese yen is worth 0.009545 U.S. Dollar, how many yen would it cost a Japanese importer for merchandise worth $550,500 from a U.S. company?

SOLUTION: 57,674,174.96

34.

If the U.S. Dollar is worth 105.16 yen, how much would a tourist using U.S. dollars pay for an airline ticket costing 478,500 yen?

SOLUTION: $4,550.21

LEARNING OBJECTIVE 2

36.

Global Concerns, Inc. has contracted to sell certain goods to a company in Australia. The Australian company has contracted to pay 500,000 Australian Dollars for the shipment. At the time the contract was signed, the Foreign Currency in Dollars column in the financial section of the morning paper showed that one Australian Dollar was valued at $0.5427 U.S. Dollar. Compute the value in U.S. currency Global Concerns expects to receive from the transaction.

SOLUTION: $271,350 U.S. currency

37.

Global Concerns, Inc. has contracted to sell certain goods to a company in Australia. The Australian company has contracted to pay 500,000 Australian Dollars for the shipment. At the time the contract was signed, the Foreign Currency in Dollars column in the financial section of the morning paper showed that one Australian Dollar was valued at $0.5427 U.S. currency. Between the date the contract was signed and the date on which payment was received, the Australian Dollar fell to a value of $0.50. Compute the amount Global Concerns gained or lost by agreeing to accept payment in Australian Dollars instead of U.S. Dollars.

SOLUTION: $21,350 U.S. currency gain

38.

White's Manufacturing, Inc. has contracted to sell certain goods to a company in the Netherlands. The price agreed upon for the goods is 94,000 Netherlands guilders. On the date the contract was signed, the Foreign Currency in Dollars column in the financial section of the local paper showed that the Netherlands guilder was valued at 0.4184. Compute the U.S. currency value White's Manufacturing, Inc. expects to receive for the goods.

SOLUTION: $39,330

39.

Roger's Manufacturing, Inc. has contracted to sell certain goods to a company in Saudi Arabia. The price agreed upon for the goods is 250,000 Saudi Arabian Riyals. On the date the contract was signed, the Foreign Currency in Dollars column in the financial section of the local paper showed that the Saudi Arabian Riyal was valued at .2267. If the value of the Saudi Arabian Riyal fell from .2267 to .2006 on the date of payment, compute how much Roger's Manufacturing, Inc. lost by contracting in Saudi Arabian Riyals instead of U.S. Dollars.

SOLUTION: $6,525

40.

Green's Manufacturing, Inc. has contracted to sell certain goods to a company in Norway. The price agreed upon for the goods is 150,000 Norwegian Kroners. On the date the contract was signed, the Foreign Currency in Dollars column in the financial section of the local paper showed that the Norwegian Kroner was valued at .1121. If the Norwegian Kroner rose to .1150 on the date of payment, compute how much Green's Manufacturing, Inc. gained by contracting in Norwegian Kroners instead of U.S. Dollars.

SOLUTION: $435

41.

Madison Manufacturing, Inc. has contracted to sell certain goods to a company in Malaysia. The price agreed upon for the goods is 320,000 Malaysian Ringgits. On the date the contract was signed, the Foreign Currency in Dollars column in the financial section of the local paper showed that the Malaysian Ringgit was valued at .2632. Compute the U.S. Dollar value Madison Manufacturing, Inc. expects to receive for the goods.

SOLUTION: $84,224 U.S. Dollars

42.

Fast Manufacturing, Inc. has contracted to sell certain goods to a company in Finland. The price agreed upon for the goods is 500,000 Euro. On the date the contract was signed, the Euro was valued at $1.30. If the value of the Euro fell from $1.30 to $1.25 on the date of payment, compute how much Fast Manufacturing, Inc. lost by contracting in Euro instead of U.S. Dollars.

SOLUTION: $25,000

43.

Great Manufacturing, Inc. has contracted to sell certain goods to a company in Austria. The price agreed upon for the goods is 900,000 Euro. On the date the contract was signed, the Euro was valued at $1.29. If the value of the Euro rose from $1.29 to $1.30 on the date of payment, compute how much Great Manufacturing, Inc. gained by contracting in Euros instead of U.S. Dollars.

SOLUTION: $9,000

44.

On a certain date the U.S. Dollar is valued at 5.7757 per Danish Krone and 0.528821 per British Pound. Compute how many more Krone than Pounds a U.S. citizen would get for $100 U.S. currency. (Round to nearest whole number)

SOLUTION: 525

45.

A German firm has contracted to sell a U.S. firm goods valued at 140,000 Euro. The Euro is valued at $1.28. Compute the value of the shipment in U.S. Dollars.

SOLUTION: $179,200.

LEARNING OBJECTIVE 3

46.

Princess Jewelry, Inc. has contracted to purchase 144 bracelets from a foreign manufacturer. The price of each bracelet is $32. An ad valorem duty of 20% is charged on each bracelet. Compute the duty Princess Jewelry, Inc. will pay for the shipment.

SOLUTION: $921.60

47.

A major chain of department stores has contracted to purchase from a foreign manufacturer 200 large radios at $36 each and 150 small radios at $22 each. An ad valorem duty of 16% is charged on all radios. Compute the duty the purchaser will pay.

SOLUTION: $1,680

48.

ABC, Inc. plans to purchase 250 units of a certain component used in building computers in their U.S. factory. ABC, Inc. can purchase the components from country X at a price of $45 each plus an ad valorem duty of 38%, or from country Y at a price of $60 each plus an ad valorem duty of 3%. Compute the amount ABC, Inc. will save by purchasing from country Y.

SOLUTION: $75

49.

XYZ, Inc. plans to purchase 460 units of a certain component used in building computers in their U.S. factory. XYZ, Inc. can purchase the components from country A at a price of $42 each plus an ad valorem duty of 42%, from country B at a price of $54 each plus an ad valorem duty of 2%, or from a domestic manufacturer at a price of $51 each. They take the lowest bid. Compute the amount the 460 units will cost XYZ, Inc.

SOLUTION: $23,460

50.

A U.S. company located in a foreign trade zone imported $1,000,000 worth of goods. The duty rate on the goods is 7%. If 15% of the goods were moved into U.S. Customs territory for sale and 85% were exported for sale, compute how much the company saved by being located in a foreign trade zone.

SOLUTION: $59,500

LEARNING OBJECTIVE 4

51.

Company A purchased 14 pounds of coffee; 17 pounds of sugar; 22 pounds of flour; and 7 pounds of beans. Compute how much the total purchase will weigh in kilograms.

SOLUTION: 27.24 kilograms

52.

A bus driver drove 10,000 miles last month. Compute the amount of kilometers he drove.

SOLUTION: 16,090 kilometers

53.

A driver purchased 180 gallons of gasoline during his vacation travels. Compute the cost for gasoline used on this vacation when the price of gasoline was $0.55 per liter. (Round answer to nearest cent.)

SOLUTION: $374.72

54.

A hardware store sold 100 yards of rope. The rope cost $0.50 per meter. Compute the purchase price of the rope.

SOLUTION: $45.7055.

55.

Car A drove 650 kilometers; car B drove 715 kilometers; and car C drove 535 kilometers. Compute the total miles the three drivers traveled.

SOLUTION: 1,180 miles

56.
Convert 48 liters to pints.

SOLUTION: 101.424

57.
Convert 48 quarts to liters.

SOLUTION: 45.408

58.
Convert 92 meters to inches.

SOLUTION: 3,622.04

59.
Convert 60 inches to meters.

SOLUTION: 1.524

60.
If a person weighs 160 pounds, how many kilograms would he/she weigh?

SOLUTION: 72.64

Chapter 21   CORPORATE STOCKS

PROBLEMS

LEARNING OBJECTIVE 1

1.
McDonalds stock closed Wednesday at 26.75. The daily stock report in the local newspaper showed that on Thursday McDonalds had a high of 27, a low of 24, and closed at 26.15. Compute the amount that the net change column for Thursday will show.

SOLUTION: −0.60

2.
Robot, Inc. closed Thursday at 37. The daily stock report in the local newspaper showed that on Friday Robot, Inc. had a high of 42.64, a low of 36.72, and closed at 39.12. Compute the amount that the net change column for Thursday will show.

SOLUTION: +2.12

3.
Audrey bought 400 shares of Nelson stock at 18.86. Commission charges were $0.20 per share. Compute the total cost of the purchase.

SOLUTION: $7,624

4.
Jason bought 300 shares of Robot stock at 38. Commission charges were $0.20 per share. Compute the total cost of the purchase.

SOLUTION: $11,460

5.
Cinthia bought 200 shares of Atlantic stock at 14.70 and 100 shares of ITT stock at 18.25. Commission charges were $0.20 per share. Compute the total cost of the purchase.

SOLUTION: $4,825

6.
Ralph bought 100 shares of Meredian stock at 86; 100 shares of Lawson stock at 30.25; and 100 shares of Puritan stock at 22.50. Commission charges were $0.20 per share. Compute the total cost of the purchase.

SOLUTION: $13,935

7.
Fern sold 800 shares of Atlantic stock at 14.25. Commission charges were $0.20 per share. Compute the proceeds of the sale.

SOLUTION: $11,240

8.

Jorge sold 200 shares of Pacific stock at 21; 300 shares of Atlantic stock at 32; and 100 shares of Inland stock at 42. Commission charges were $0.20 per share. Compute the proceeds of the sale.

SOLUTION: $17,880

LEARNING OBJECTIVE 2

9.

Assume a differential rate of 12.5 cents for odd lot sales and purchases. Mark sold 120 shares of PFB stock at $74.50. Commission charges were $0.20 per share. Compute the proceeds of the sale.

SOLUTION: $8,913.50

10.

Assume a differential rate 12.5 cents for odd lot sales and purchases. June purchased 150 shares of UNF stock at 23. Commission charges were $0.20 per share. Compute the total purchase costs.

SOLUTION: $3,486.25

LEARNING OBJECTIVE 3

11.

Stock of the ABC Corporation has a par value of $30. If the corporation pays a 6% dividend, compute the amount of the dividend per share.

SOLUTION: $1.80

12.

Stock of the XYZ Corporation has a par value of $50. If the corporation pays a 5% dividend, compute the amount of the dividend paid to the owner of 160 shares.

SOLUTION: $400

13.

Stock of the Boston Corporation has a par value of $20. If the corporation pays a 7% dividend, compute the amount of the dividend paid to the owner of 762 shares.

SOLUTION: $1,066.80

14.

Stock of the Lynch Corporation has a par value of $35. If the corporation pays a 9% dividend, compute the amount of the dividend paid to the owner of 2,500 shares.

SOLUTION: 7,875

15.
Margaret Jones bought 1,000 shares of DVG stock at 26. Commission charges were $0.20 per share. A dividend of $1.10 per share was paid this year. Compute the rate of yield. (Round answer to two decimal places.)

SOLUTION: 4.20%

16.
Alan Jackson bought 300 shares of MTY stock at 47. Commission charges were $0.20 per share. A dividend of $1.70 per share was paid this year. Compute the rate of yield. (Round answer to two decimal places.)

SOLUTION: 3.60%

17.
Marge Nelson bought 400 shares of BRN stock at 34. Commission charges were $0.20 per share. A dividend of $2.20 per share was paid this year. Compute the rate of yield. (Round answer to two decimal places.)

SOLUTION: 6.43%

18.
George Fuji bought 600 shares of TTV stock at 28. Commission charges were $0.20 per share. A dividend of $1.50 per share was paid this year. Compute the rate of yield. (Round answer to two decimal places.)

SOLUTION: 5.32%

19.
Mildred Smith bought 400 shares of DVG stock at 21. She sold the stock at 32. Commission charges were $0.20 per share. Compute the dollar amount of gain or loss.

SOLUTION: $4,240 (gain)

20.
Sam Cooper bought 200 shares of MTV stock at 54. He sold the stock at 52. Commission charges were $0.20 per share. Compute the dollar amount of gain or loss.

SOLUTION: −480 (loss)

21.
Marge Nelson bought 300 shares of BRN stock at 34. She sold the stock at 35. Commission charges were $0.20 per share. Compute the dollar amount of gain or loss.

SOLUTION: $180 (gain)

22.
George Fuji bought 400 shares of TTV stock at 28. He sold the stock at 34. Commission charges were $0.20 per share. Compute the dollar amount of gain or loss.

SOLUTION: $2,240 (gain)

LEARNING OBJECTIVE 4

23.

The Channing Company earned $64,000 last year. The capital stock of the company consists of $500,000 of 8% preferred stock and $200,000 of common stock. If the directors declare a dividend of the entire earnings, compute the total amount that will be paid to the holders of preferred stock.

SOLUTION: $40,000

24.

The Channing Company earned $64,000 last year. The capital stock of the company consists of $500,000 of 8% preferred stock and $200,000 of common stock. If the directors declare a dividend of the entire earnings, compute the total amount that will be paid to the holders of common stock.

SOLUTION: $24,000

25.

The Seabrite Company earned $84,000 last year. The capital stock of the company consists of $500,000 of 6% preferred stock and $800,000 of common stock. If the directors declare a dividend of the entire earnings, compute the total amount that will be paid to the holders of preferred stock.

SOLUTION: $30,000

26.

The Seabrite Company earned $84,000 last year. The capital stock of the company consists of $500,000 of 6% preferred stock and $800,000 of common stock. If the directors declare a dividend of the entire earnings, compute the total amount that will be paid to the holders of common stock.

SOLUTION: $54,000

27.

The Swartz Company earned $30,000 last year. The capital stock of the company consists of $200,000 of 7% preferred stock and $100,000 of common stock. If the directors declare a dividend of half of the entire earnings, compute the total amount that will be paid to the holders of preferred stock.

SOLUTION: $14,000

28.

The Swartz Company earned $30,000 last year. The capital stock of the company consists of $200,000 of 7% preferred stock and $100,000 of common stock. If the directors declare a dividend of half of the entire earnings, compute the total amount that will be paid to the holders of common stock.

SOLUTION: $1,000

29.

The Scott Company earned $120,000 last year. The capital stock of the company consists of $400,000 of 7% preferred stock and $200,000 of common stock. If the directors declared a dividend of 60% of the earnings, compute the total amount that will be paid to the holders of preferred stock.

SOLUTION: $28,000

30.

The Scott Company earned $120,000 last year. The capital stock of the company consists of $400,000 of 7% preferred stock and $200,000 of common stock. If the directors declared a dividend of 60% of the earnings, compute the total amount that will be paid to the holders of common stock.

SOLUTION: $44,000

31.

The Hybrid Company earned $52,000 last year. The capital stock of the company consists of $300,000 of 7% cumulative preferred stock and $200,000 of common stock. The directors declared a dividend of the entire earnings. During the previous year, the company earned only enough to pay a 2% dividend on preferred stock. What is the total amount that will be paid to the holders of preferred stock?

SOLUTION: $36,000

32.

The Hybrid Company earned $52,000 last year. The capital stock of the company consists of $300,000 of 7% cumulative preferred stock and $200,000 of common stock. The directors declared a dividend of the entire earnings. During the previous year, the company earned only enough to pay a 2% dividend on preferred stock. What is the total amount that will be paid to the holders of common stock?

SOLUTION: $16,000

33.

The Tallman Company earned $120,000 last year. The capital stock of the company consists of $700,000 of 6% cumulative preferred stock and $1,500,000 of common stock. The directors declared a dividend of the entire earnings. During the previous year, the company earned only enough to pay a 3% dividend on preferred stock. What is the total amount that will be paid to the holders of preferred stock?

SOLUTION: $63,000

34.

The Tallman Company earned $120,000 last year. The capital stock of the company consists of $700,000 of 6% cumulative preferred stock and $1,500,000 of common stock. The directors declared a dividend of the entire earnings. During the previous year, the company earned only enough to pay a 3% dividend on preferred stock. What is the total amount that will be paid to the holders of common stock?

SOLUTION: $57,000

35.

The Pratt Company earned $140,000 last year. The capital stock of the company consists of $600,000 of 8% cumulative preferred stock and $600,000 of common stock. The directors declared a dividend of half of the earnings. During the previous year, the company earned only enough to pay a 6% dividend on preferred stock. What is the total amount that will be paid to the holders of preferred stock?

SOLUTION: $60,000

36.

The Pratt Company earned $140,000 last year. The capital stock of the company consists of $600,000 of 8% cumulative preferred stock and $600,000 of common stock. The directors declared a dividend of half of the earnings. During the previous year, the company earned only enough to pay a 6% dividend on preferred stock. Compute the total amount that will be paid to the holders of common stock?

SOLUTION: $10,000

37.

The Buckley Company earned $110,000 last year. The capital stock of the company consists of $300,000 of 9% cumulative preferred stock and $200,000 of common stock. The directors declared a dividend of 60% of the earnings. For the previous year, the directors did not declare a dividend at all. Compute the total amount that will be paid to the holders of preferred stock.

SOLUTION: $54,000

38.

The Buckley Company earned $110,000 last year. The capital stock of the company consists of $300,000 of 9% cumulative preferred stock and $200,000 of common stock. The directors declared a dividend of 60% of the earnings. For the previous year, the directors did not declare a dividend at all. Compute the total amount that will be paid to the holders of common stock.

SOLUTION: $12,000

39.

Christopher Cooper owned 135 shares of Weaver Company's convertible preferred stock at $40 par value. He converted each share of preferred stock into two shares of common stock. Compute the number of shares of common stock Christopher Cooper received when he converted.

SOLUTION: 270

40.

Christopher Cooper owned 135 shares of Weaver Company's convertible preferred stock at $40 par value. He converted each share of preferred stock into two shares of common stock. If common stock was selling at $26 per share on the date of conversion, compute Christopher Cooper's common stock worth.

SOLUTION: $7,020

41.

Christopher Cooper owned 135 shares of Weaver Company's convertible preferred stock at $40 par value. He converted each share of preferred stock into two shares of common stock. If Christopher Cooper paid $40 per share for his preferred stock, and if common stock was selling at $26 per share on the date of conversion, compute the amount that Christopher Cooper's investment increased in value.

SOLUTION: $1,620

42.

Patrick Harrigan owned 200 shares of Marsh Company's convertible preferred stock at $12 par value. He converted each share of preferred stock into two shares of common stock. If the convertible preferred stock paid 8% annually and the common stock pays $0.60 a share this year, how much more dividend will Patrick Harrigan receive this year because of his conversion.

SOLUTION: $48

43.

Mary Lee owned 500 shares of Dugan Company's convertible preferred stock at $20 par value. She converted each share of preferred stock into four shares of common stock. Compute the number of shares of common stock that Mary Lee received when she converted.

SOLUTION: 2,000

44.

Mary Lee owned 500 shares of Dugan Company's convertible preferred stock at $20 par value. She converted each share of preferred stock into four shares of common stock. If common stock was selling at $50 per share on the date of conversion, compute Mary Lee's common stock worth.

SOLUTION: $100,000

45.

Mary Lee owned 500 shares of Dugan Company's convertible preferred stock at $20 par value. She converted each share of preferred stock into four shares of common stock. If Mary Lee paid $20 per share for her preferred stock, and if common stock was selling at $50 per share on the date of conversion, compute the amount that Mary Lee's investment increased in value.

SOLUTION: $90,000

46.

Mary Lee owned 500 shares of Dugan Company's convertible preferred stock at $20 par value. She converted each share of preferred stock into four shares of common stock. If the convertible preferred stock paid 9% annually and the common stock pays $1.00 a share this year, how much more dividend will Mary Lee receive this year because of her conversion?

SOLUTION: $1,100

47.

Bob Wright owned 100 shares of Lawson Company's convertible preferred stock at $10 par value. He converted each share of preferred stock into three shares of common stock. Compute the number of shares of common stock Bob Wright received when he converted.

SOLUTION: 300

48.

Bob Wright owned 100 shares of Lawson Company's convertible preferred stock at $10 par value. He converted each share of preferred stock into three shares of common stock. If common stock was selling at $4 per share on the date of conversion, compute Bob Wright's common stock worth.

SOLUTION: $1,200

49.

Bob Wright owned 100 shares of Lawson Company's convertible preferred stock at $10 par value. He converted each share of preferred stock into three shares of common stock. If Bob Wright paid $10 per share for his preferred stock, and if common stock was selling at $4 per share on the date of conversion, compute the amount that Bob Wright's investment increased in value.

SOLUTION: $200

50.

Bob Wright owned 100 shares of Lawson Company's convertible preferred stock at $10 par value. He converted each share of preferred stock into three shares of common stock. If the convertible preferred stock paid 7% annually and the common stock pays $0.40 a share this year, how much more dividend will Bob Wright receive this year because of his conversion?

SOLUTION: $50

LEARNING OBJECTIVES 1-4

51 – 54.

Use the following stock listing to answer Questions 51 through 56.

| 52-Week | | | | Vol. | | | | |
Hi	Lo	Stock	Div.	100s	Hi	Lo	Close	Chg.
32	27.5	EXN	1.45	85	30.62	29.5	30.25	+1.2

51.

What is the PE ratio if EXN has earnings per share of $1.89?

SOLUTION: 16

52.

How many shares were sold?

SOLUTION: 8,500

53.

What was the closing price of the previous day?

SOLUTION: $29.05

54.

What is the rate of yield for the EXN stock? Round to nearest tenth of a percent.

SOLUTION: 4.8%

55.

John purchased 400 shares of MMM stock at 78. One year later he sold all 400 shares at 82. He paid a transaction fee of $19.95 for each transaction. What was his gain or loss on the sale?

SOLUTION: $1,560.10 gain

56.

May purchased 500 shares of XYZ stock at 31.45. One year later she sold the 500 shares at 26.97. She paid a transaction fee of $19.95 for each transaction. What was her gain or loss on the sale?

SOLUTION: $2,279.90 Loss

57.

Beta Corporation's capital consists of 15,000 shares of $50 par 7% preferred stock and 50,000 shares of no-par common stock. The board of directors declared a dividend of $112,500. Calculate the dividend per share for preferred stock.

SOLUTION: $3.50

58.

Beta Corporation's capital consists of 15,000 shares of $50 par 7% preferred stock and 50,000 shares of no-par common stock. The board of directors declared a dividend of $112,500. Calculate the dividend per share for common stock.

SOLUTION: $1.20

59.

Marge Brown bought 200 shares of Kenworth $50 par 7% preferred stock at the current market price of $58 per share. What is her rate of yield if Kenworth pays the regular dividend for the year?

SOLUTION: 6.03%

60.

Beth Martin bought 300 shares of Alpha $50 par 6% preferred stock at the current market price of $59 per share. What is her rate of yield if Alpha pays the regular dividend for the year?

SOLUTION: 5.08%

Chapter 22 CORPORATE AND GOVERNMENT BONDS

PROBLEMS

LEARNING OBJECTIVE 1

1.

Sandia Corporation issued $2,000,000 worth of callable bonds paying 7% interest. The maturity date for the bonds was in 10 years. A year later, interest rates fell to 5%. The bonds were called and new bonds were sold at the 5% rate. How much did Worldwide Corporation save by calling the bonds?

SOLUTION: $360,000

2.

Erik Wells bought a Sandia Corporation convertible bond for $1,000. The bond was convertible to 50 shares of stock. At the time of the purchase, the stock was selling for $20 per share. At the end of two years, the stock was selling for $28 per share. Mark converted. Assuming the market value of the bond had not changed, how much profit did Erik realize by converting?

SOLUTION: $400

3.

What would be the "stock" value of a bond that was convertible to 30 shares of stock if the stock was priced at 38.40?

SOLUTION: $1,152

4.

What would be the "stock" value of a bond that was convertible to 50 shares of stock if the stock was priced at 27.32?

SOLUTION: $1,366

5.

If a company issued a callable bond at 6% interest, would it be likely to call the bond if the current rate of interest rose to 7%?

SOLUTION: no

6.

Allison Yu purchased five $1,000 convertible bonds at face value. Each bond was convertible into 25 shares of common stock. After several years, when the stock was selling at 46, Allison converted all five bonds. What was Allison's gain upon conversion of the bonds?

SOLUTION: $750

LEARNING OBJECTIVE 2

7.
Compute the dollar amount of interest that will be earned per year for a bond listed as Raz9¼s12.

SOLUTION: $92.50

8.
Compute the dollar amount of interest that will be earned per year for a bond listed as RKB8¾s15.

SOLUTION: $87.50

9.
Compute the dollar amount of interest that will be earned per year for one bond listed as Marr8-3/8s17 and one bond listed as Pudt8-7/8s09.

SOLUTION: $172.50

10.
Compute the dollar amount of interest that will be earned per year for three bonds listed as Hertz7.2s13 and two bonds listed as Cmz9s18.

SOLUTION: $396

11.
Compute the dollar amount of interest that will be earned per year for four bonds listed as Kvr7¾s12 and one bond listed as LMC8s17.

SOLUTION: $390

12.
Compute the dollar amount of interest that will be earned per year for three bonds listed as ATT10s18 and two bonds listed as IBM9½s20.

SOLUTION: $490

13.
Compute the discounted price at which a $1,000 bond quoted at 87¼ would sell.

SOLUTION: $872.50

14.
Compute the premium price at which a $1,000 bond quoted at 108½ would sell.

SOLUTION: $1,085

15.
Is a bond quoted at 90 selling at a premium or a discount?

SOLUTION: Discount

16.

Is a bond quoted at 105 selling at a premium or a discount?

SOLUTION: Premium

LEARNING OBJECTIVE 3

17.

A $1,000 bond with interest at 9 1/2% on March 1 and September 1 was purchased on November 23. Compute the number of days for which accrued interest will be paid.

SOLUTION: 83days

18.

A $1,000 bond with interest at 6 7/8% on January 1 and July 1 was purchased on October 7. Compute the number of days for which accrued interest will be paid.

SOLUTION: 98 days

19.

A $1,000 bond with interest at 9 1/8% on March 1 and September 1 was purchased on October 15. Compute the number of days for which accrued interest will be paid.

SOLUTION: 44 days

20.

A $1,000 bond with interest at 6.4% paid semiannually January 1 and July 1 was purchased on August 28. Compute the number of days for which accrued interest will be paid.

SOLUTION: 58 days

21.

A $1,000 bond with interest at 8% on March 1 and September 1 was purchased on October 17. Compute the dollar amount of accrued interest that will be paid to the seller. (Assume a 360-day year.)

SOLUTION: $10.22

22.

A $1,000 bond with interest at 8% on March 1 and September 1 was sold on
July 8 at 92 plus accrued interest. Compute the dollar amount of the sale the seller received. (Assume a 360-day year and a commission of $5 per bond.)

SOLUTION: $943.66

23.

A $1,000 bond with interest at 9% on January 1 and July 1 was purchased on September 12 at 86 plus accrued interest. Compute the entire purchase cost of the bond. (Assume a 360-day year and a commission of $5 per bond.)

SOLUTION: $883.25

24.
A $1,000 bond with interest at 9% on January 1 and July 1 was sold on March 20 at 109 plus accrued interest. Compute the dollar amount of the sale the seller received. (Assume a 360-day year and that it is not a leap year and a commission of $5 per bond.)

SOLUTION: $1,104.50

25.
A $1,000 bond with interest at 8 1/2% on March 1 and September 1 was purchased on June 18 at 108 plus accrued interest. Compute the entire purchase cost of the bond. (Assume a 360-day year and a commission of $5 per bond.)

SOLUTION: $1,110.24

26.
A $1,000 bond with interest at 9 1/2% on March 1 and September 1 was sold on July 7 at 102 plus accrued interest. Compute the dollar amount of the sale the seller received. (Assume a 360-day year and a commission of $5 per bond.)

SOLUTION: $1,048.78

27.
A $1,000 bond with interest at 9 1/4% on January 1 and July 1 was purchased on September 10. Compute the dollar amount of accrued interest that will be paid to the seller. (Assume a 360-day year. Round answer to nearest cent.)

SOLUTION: $18.24

28.
A $1,000 bond with interest at 10% on March 1 and September 1 was purchased on February 13. Compute the dollar amount of accrued interest that will be paid to the seller. (Assume a 360-day year. Round answer to nearest cent.)

SOLUTION: $45.83

29.
A $1,000 bond with interest at 7% on January 1 and July 1 was purchased on May 11. Compute the dollar amount of accrued interest that will be paid to the seller. (Assume a 360-day year and that it is not a leap year. Round answer to nearest cent.)

SOLUTION: $25.28

30.
A $1,000 bond with interest at 9% on March 1 and September 1 was purchased on November 4 at 107 plus accrued interest. Compute the entire purchase cost of the bond. (Assume a 360-day year and a commission of $5 per bond.)

SOLUTION: $1,091

31.

A $1,000 bond with interest at 9% on March 1 and September 1 was sold on October 30 at 105 plus accrued interest. Compute the dollar amount of the sale the seller received. (Assume a 360-day year and a commission of $5 per bond.)

SOLUTION: $1,059.75

32.

A $1,000 bond with interest at 8 1/2% on January 1 and July 1 was purchased on September 30 at 97 plus accrued interest. Compute the entire purchase cost of the bond. (Assume a 360-day year and a commission of $5 per bond.)

SOLUTION: $996.49

33.

A $1,000 bond with interest at 8 1/2% on January 1 and July 1 was sold on September 4 at 103 plus accrued interest. Compute the dollar amount of the sale the seller received. (Assume a 360-day year and a commission of $5 per bond.)

SOLUTION: $1,040.35

34.

A $1,000 bond with interest at 8% on March 1 and September 1 was purchased on October 8 at 104 plus accrued interest. Compute the entire purchase cost of the bond. (Assume a 360-day year and a commission of $5 per bond.)

SOLUTION: $1,053.22

35.

On November 15, Melvin Weldon purchased five KTV8s12 bonds that pay interest semiannually April 1 and October 1. Calculate the accrued interest Melvin paid to the seller. (Assume a 360-day year.)

SOLUTION: $50

36.

On June 20, Betty Carper purchased eight JVT7.2s09 bonds that pay interest semiannually April 1 and October 1. Calculate the accrued interest Betty paid to the seller. (Assume a 360-day year.)

SOLUTION: $128

37.

Five $1,000 bonds that pay interest at 9% semiannually April 1 and October 1 were purchased July 10 at 92. Calculate the total amount paid for the bonds including accrued interest and commission of $5 per bond. (Use a 360-day year.)

SOLUTION: $4,750

LEARNING OBJECTIVES 4, 5

TEXT, PAGE 6 USES CURRENT YIELD RATHER THAN AVERAGE ANNUAL YIELD

38.
An investor bought an 8% bond at 106. The bond would mature in 5 years. Compute the rate of yield to maturity. (Do not consider commission. Round answer to two decimal places.)

SOLUTION: 6.60%

39.
An investor bought a 9% bond at 88. The commission was $5. Compute the current yield. (Round answer to two decimal places.)

SOLUTION: 10.17%
The text examples do not mention commissions when covering current yield.

40.
An investor bought an 8 1/2% bond at 108. Compute the current yield. (Round answer to two decimal places.)

SOLUTION: 7.87%

41.
An investor bought an 8 1/2% bond at 108. The bond would mature in 8 years. Compute the rate of yield to maturity. (Round answer to two decimal places.)

SOLUTION: 7.21%

42.
An investor bought a 7% bond at 82. Compute the current yield. (Round answer to two decimal places.)

SOLUTION: 8.53%

43.
An investor bought a 7% bond at 82. The bond would mature in 6 years. Compute the rate of yield to maturity. (Do not consider commission. Round answer to two decimal places.)

SOLUTION: 10.99%

44.
An investor bought a 10 1/2% bond at 109. The commission was $5. Compute the current yield. (Round answer to two decimal places.)

SOLUTION: 9.59%

45.
An investor bought a 10 1/2% bond at 109. The bond would mature in 5 years. Compute the rate of yield to maturity. (Do not consider commission. Round answer to two decimal places.)

SOLUTION: 8.33%

46.
An investor bought a 6 3/4% bond at 74. The commission was $5. Compute the current yield. (Round answer to two decimal places.)

SOLUTION: 9.06%

47.
An investor bought a 6 3/4% bond at 74. The bond would mature in 13 years. Compute the rate of yield to maturity. (Do not consider commission. Round answer to two decimal places.)

SOLUTION: 10.06%

48.
An investor bought an 11% bond at 112. The commission was $5. Compute the current yield. (Round answer to two decimal places.)

SOLUTION: 9.78%

49.
An investor bought an 11% bond at 112. The bond would mature in 10 years. Compute the rate of yield to maturity. (Do not consider commission. Round answer to two decimal places.)

SOLUTION: 9.25%

50.
An investor bought two bonds. Bond A was a 9% bond at 106 and bond B was a 7% bond at 94. Commission was $5 per bond. Compute how much greater is the current yield from bond A is than from bond B. (Round yields to two decimal places.)

SOLUTION: 1.04% greater yield from Bond A

51.
An investor bought two bonds. Bond A was a 9% bond at 106 and bond B was a 7% bond at 94. Each bond is to mature in 5 years. Compute how much greater the yield to maturity from bond B is than from bond A. (Do not consider commission. Round yields to two decimal places.)

SOLUTION: 0.88% greater yield from Bond B

52.
An investor bought two bonds. Bond A was an 8% bond at 102 and bond B was an 8 1/2% bond at 104. Commission was $5 per bond. Compute how much greater the current yield from bond B is than from bond A. (Round yields to two decimal places.)

SOLUTION: 0.33% greater yield from Bond B

53.
An investor bought two bonds. Bond A was an 8% bond at 102 and bond B was an 8 1/2% bond at 104. Each bond is to mature in 4 years. Compute how much greater the yield to maturity from bond A is than from bond B. (Do not consider commission. Round yields to two decimal places.)

SOLUTION: 0.08% greater yield from Bond A

54.

An investor bought a 5 1/2% bond at 99. Compute the current yield. (Round answer to two decimal places.)

SOLUTION: 5.55%

55.

An investor bought an 8 1/2% bond at 98. The bond would mature in 4 years. Compute the rate of yield to maturity. (Do not consider commission. Round answer to two decimal places.)

SOLUTION: 9.09%

56.

An investor bought a 9% bond at 88. The bond would mature in 6 years. Compute the rate of yield to maturity. (Do not consider commission. Round answer to two decimal places.)

SOLUTION: 11.70%

57.

An investor purchased a 6.4% bond at 92. The bond matures in 8 years. Calculate the current yield. (Round answer to two decimal places.)

SOLUTION: 6.96%

58.

An investor purchased a 7% bond at 90 (no commission). The bond matures in 10 years. Calculate the rate of yield to maturity. (Round answer to two decimal places.)

SOLUTION: 8.42%

59.

An investor purchased an 8½% bond at 108. The bond matures in 5 years. Calculate the current yield. (Round answer to two decimal places.)

SOLUTION: 7.87%

Chapter 23 ANNUITIES

Note 1: Tables 23-1A, 23-1B, 23-2A, and 23-2B appear at the end of this chapter.
Note 2: If a calculator is used to find an annuity factor, the answer may vary slightly.

PROBLEMS

LEARNING OBJECTIVES 1, 2

1.

Each of the following annuities involves future values. Compute the missing numbers. Use Tables 23-1A and 23-1B or a calculator.

	Payment Amount	Payment Periods	Interest Rate	Length of Annuity	Future Value
a.	_____	quarterly	6% compounded quarterly	10 years	$64,300
b.	$3,250	monthly	6% compounded monthly	3 years	_____

SOLUTION:
a. $64,300 ÷ 54.26789 = $1,184.86 Payment Amount
b. $3,250 × 39.33610 = $127,842.33 Future Value

2.

Each of the following annuities involves future values. Compute the missing numbers. Use Tables 23-1A and 23-1B or a calculator.

	Payment Amount	Payment Periods	Interest Rate	Length of Annuity	Future Value
a.	$25,000	annually	6% compounded annually	4 years	_____
b.	_____	semiannually	6% compounded semiannually	9 years	$40,000

SOLUTION:
a. $25,000 × 4.37462 = $109,365.50 Future Value
b. $40,000 ÷ 23.41444 = $1,708.35 Payment Amount

3.

Each of the following annuities involves future values. Compute the missing numbers. Use Tables 23-1A and 23-1B or a calculator.

	Payment Amount	Payment Periods	Interest Rate	Length of Annuity	Future Value
a.	_____	annually	10% compounded annually	7 years	$ 25,000
b.	$12,500	quarterly	6% compounded quarterly	5 years	_____

SOLUTION:
a. $25,000 ÷ 9.48717 = $2,635.14 Payment Amount
b. $12,500 × 23.12367 = $289,045.88 Future Value

4.
Each of the following annuities involves future values. Compute the missing numbers. Use Tables 23-1A and 23-1B or a calculator.

	Payment Amount	Payment Periods	Interest Rate	Length of Annuity	Future Value
a.	$ 1,200	monthly	12% compounded monthly	2 years	_____
b.	_____	quarterly	8% compounded quarterly	7 years	$ 7,900

SOLUTION:
a. $1,200 × 26.97346 = $32,368.15 Future Value
b. $7,900 ÷ 37.05121 = $213.22 Payment Amount

5.
Each of the following annuities involves future values. Compute the missing numbers. Use Tables 23-1A and 23-1B or a calculator.

	Payment Amount	Payment Periods	Interest Rate	Length of Annuity	Future Value
a.	_____	quarterly	12% compounded quarterly	30 months	$ 3,800
b.	$1,600	monthly	15% compounded monthly	1.5 years	_____

SOLUTION:
a. $3,800 ÷ 11.46388 = $331.48 Payment Amount
b. $1,600 × 20.04619 = $32,073.91 Future Value

6.
Each of the following annuities involves future values. Compute the missing numbers. Use Tables 23-1A and 23-1B or a calculator.

	Payment Amount	Payment Periods	Interest Rate	Length of Annuity	Future Value
a.	_____	monthly	12% compounded monthly	2 years	$100,000
b.	$6,000	semiannually	10% compounded semiannually	13 years	_____

SOLUTION:
a. $100,000 ÷ 26.97346 = $3,707.35 Payment Amount
b. $6,000 × 51.11345 = $306,680.70 Future Value

LEARNING OBJECTIVE 1

7.
Brenda Davies inherited a poultry company from her uncle. To make necessary improvements, she began to set aside $6,000 every three months. Brenda earns 8% compounded quarterly. Compute the total amount that she will have after 45 months. Use Tables 23-1A and 23-1B or a calculator.

SOLUTION:
$6,000 × 17.29342 = $103,760.52 Future value

8.

Nick Yeager, an accountant, will retire in 11 years. Nick invests $10,000 every year into a fund that promises to return 9% compounded annually. Compute the total amount that he will have at the end of 11 years. Use Tables 23-1A and 23-1B or a calculator.

SOLUTION:
$10,000 × 17.56029 = $175,602.90 Future value

LEARNING OBJECTIVE 2

9.

Amit Cheeda has a welding business that services primarily heavy equipment on highway construction projects. Amit plans to replace equipment in five years but he does not want to borrow the money. Therefore, he makes twenty quarterly deposits into a sinking fund that will earn 6% compounded quarterly. How much should Amit deposit each quarter if he wants to have $50,000 after the five years? Use Tables 23-1A and 23-1B or a calculator.

SOLUTION:
$50,000 ÷ 23.12367 = $2,162.29 Quarterly payments

LEARNING OBJECTIVE 1

10.

Carolyn Green is a single parent who plans to save for her child's education. She can afford to save $250 every month for 2 1/2 years. If Carolyn could interest of 12% compounded monthly, compute the total amount that she will accumulate during the 2 1/2 years. Use Tables 23-1A and 23-1B or a calculator.

SOLUTION:
$250 × 34.78489 = $8,696.22 Future value

LEARNING OBJECTIVE 1

11.

$375 is invested each month for 2 years. Compute the future value if interest is 12% compounded monthly. Use Tables 23-1A and 23-1B or a calculator.

SOLUTION:
$375 × 26.97346 = $10,115.05 Future value

12.

$1,125 is invested each quarter for 11 years. Compute the future value if interest is 8% compounded quarterly. Use Tables 23-1A and 23-1B or a calculator.

SOLUTION:
$1,125 × 69.50266 = $78,190.49 Future value

LEARNING OBJECTIVE 2

13.
Compute the size of deposit that is required every six months to have a future value of $6,000 at the end of 4 years if the interest rate is 10% compounded semiannually. Use Tables 23-1A and 23-1B or a calculator.

SOLUTION:
$6,000 ÷ 9.54911 = $628.33 Deposit required

14.
Compute the size of deposit that is required every year to have a future value of $37,600 at the end of 23 years if the interest rate is 5% compounded annually. Use Tables 23-1A and 23-1B or a calculator.

SOLUTION:
$37,600 ÷ 41.43048 = $907.54 Deposit required

15.
Leslie Yu wants to deposit an equal amount each month into an account that will pay 6% compounded monthly. If Leslie wants to end up with a future value of $5,000 after 20 months, how much should she deposit each month? Use Tables 23-1A and 23-1B or a calculator.

SOLUTION:
$5,000 ÷ 20.97912 = $238.33 Deposit required

LEARNING OBJECTIVE 1

16.
Compute the future value in three years if $1,925 is invested every three months into a project that pays 16% compounded quarterly. Use Tables 23-1A and 23-1B or a calculator.

SOLUTION:
$1,925 × 15.02581 = $28,924.68 Future value

LEARNING OBJECTIVE 2

17.
An investment pays 9% compounded annually. Compute the amount that must be invested each year for five years to have a total accumulation of $8,000. Use Tables 23-1A and 23-1B or a calculator.

SOLUTION:
$8,000 ÷ 5.98471 = $1,336.74 Invested each year

LEARNING OBJECTIVE 1

18.

Compute the total that will accumulate in 6 years if $250 is deposited every quarter into a bank account that pays 8% compounded quarterly? Use Tables 23-1A and 23-1B or a calculator.

SOLUTION:
$250 × 30.42186 = $7,605.47 Future value

19.

Jacklyn Avery made two $600 deposits every year (i.e., semiannual) for 7 years. If the investment pays a return of 10% compounded semiannually, how much would Jacklyn's investment earn during the seven years? Use Tables 23-1A and 23-1B or a calculator.

SOLUTION:
$600 × 19.59863 = $11,759.18; $600 × 14 = $8,400;
$11,759.18 – $8,400 = $3,359.18 Interest earned

LEARNING OBJECTIVE 2

20.

Eliberto Ochoa needs to have $7,500 at the end of 15 months to pay for a new roof. Compute the amount that Eliberto should deposit each month into a sinking fund that pays 9% compounded monthly. Use Tables 23-1A and 23-1B or a calculator.

SOLUTION:
$7,500 ÷ 15.81368 = $474.27 Deposited each month

LEARNING OBJECTIVES 3, 4

21.

Each of the following annuities involves present values. Compute the missing numbers. Use Tables 23-2A and 23-2B or a calculator.

	Payment Amount	Payment Periods	Interest Rate	Length of Annuity	Future Value
a.	$15,000	annually	5% compounded annually	17 yrs	_____
b.	_____	semiannually	6% compounded semiannually	19 yrs	$28,200

SOLUTION:
a. $15,000 × 11.27407 = $169,111.05 Present Value
b. $28,200 ÷ 22.49296 = $1,253.75 Payment Amount

22.

Each of the following annuities involves present values. Compute the missing numbers. Use Tables 23-2A and 23-2B or a calculator.

	Payment Amount	Payment Periods	Interest Rate	Length of Annuity	Future Value
a.	$ 1,000	monthly	6% compounded monthly	2 1/2 yrs	_____
b.	_____	quarterly	6% compounded quarterly	42 months	$ 8,700

SOLUTION:
a. $1,000 × 27.79405 = $27,794.05 Present Value
b. $8,700 ÷ 12.54338 = $693.59 Payment Amount

23.

Each of the following annuities involves present values. Compute the missing numbers. Use Tables 23-2A and 23-2B or a calculator.

	Payment Amount	Payment Periods	Interest Rate	Length of Annuity	Future Value
a.	_____	semiannually	10% compounded semiannually	42 months	$30,000
b.	$ 8,000 annually		5% compounded annually	17 years	_____

SOLUTION:
a. $30,000 ÷ 5.78637 = $5,184.60 Payment Amount
b. $8,000 × 11.27407 = $90,192.56 Present Value

24.

Each of the following annuities involves present values. Compute the missing numbers. Use Tables 23-2A and 23-2B or a calculator.

	Payment Amount	Payment Periods	Interest Rate	Length of Annuity	Future Value
a.	_____	monthly	6% compounded monthly	1 1/4 yrs	$17,800
b.	$ 2,500	quarterly	12% compounded quarterly	6 years	_____

SOLUTION:
a. $17,800 ÷ 14.41662 = $1,234.69 Payment Amount
b. $2,500 ×16.93554 = $42,338.85 Present Value

25.

Each of the following annuities involves present values. Compute the missing numbers. Use Tables 23-2A and 23-2B or a calculator.

	Payment Amount	Payment Periods	Interest Rate	Length of Annuity	Future Value
a.	$ 8,800	annually	6% compounded annually	13 years	_____
b.	_____	quarterly	12% compounded quarterly	5 years	$25,000

SOLUTION:
a. $8,800 × 8.85268 = $77,903.58 Present Value
b. $25,000 ÷ 14.87747 = $1,680.39 Payment Amount

26.

Each of the following annuities involves present values. Compute the missing numbers. Use Tables 23-2A and 23-2B or a calculator.

	Payment Amount	Payment Periods	Interest Rate	Length of Annuity	Future Value
a.	$3,150	semiannually	12% compounded semiannually	14 years	_____
b.	_____	monthly	15% compounded monthly	1 year	$6,500

SOLUTION:
a. $3,150 × 13.40616 = $42,229.40 Present Value
b. $6,500 ÷ 11.07931 = $586.68 Payment Amount

LEARNING OBJECTIVE 3

27.

Benoit Asset Management guarantees a minimum return of 6% compounded monthly for two years. What is the total present value of twenty-four annual investment deposits of $650 each? Use Tables 23-2A and 23-2B or a calculator.

SOLUTION:
$650 × 22.56287 = $14,665.87 Present value

LEARNING OBJECTIVE 4

28.

Edgar Goloc deposited $25,000 (present value) today at 6% compounded quarterly. Compute the quarterly withdrawal that Edgar can make each quarter for 11 years and empty the account. Use Tables 23-2A and 23-2B or a calculator.

SOLUTION:
$25,000 ÷ 32.04062 = $780.26 Monthly payments

LEARNING OBJECTIVE 3

29.

From her retirement fund, Gretchen Yasui will receive $3,000 each quarter for 5 years. If Gretchen gets an interest rate of 8% compounded quarterly, compute the present value of these payments. Use Tables 23-2A and 23-2B or a calculator.

SOLUTION:
$3,000 × 16.35143 = $49,054.29 Present value

30.

Rick Rios has an asphalt paving business. He is trying to calculate when he should replace some of his current equipment. Rick estimates that he will spend $8,000 every six months over the next four years to maintain the current equipment. At a rate of 8% compounded semiannually, what is the total present value of the estimated maintenance payments? Use Tables 23-2A and 23-2B or a calculator.

SOLUTION:
$8,000 × 6.73274 = $53,861.92 Present value

31.
For the next two years, Walter and Beatrice Winetrout want to send $500 per month to their daughter in college. To make these payments, compute the amount that the Winetrouts should deposit today into an investment that will return 12% compounded monthly. Use Tables 23-2A and 23-2B or a calculator.

SOLUTION:
$500 × 21.24339 = $10,621.70 Deposit today

LEARNING OBJECTIVE 4

32.
Frances Koo has $36,700 in an investment account that pays her interest at 8% compounded semiannually. Frances will withdraw a specific amount (always the same) every six months for 12 years. After the last withdrawal, the account will be empty. Compute the amount that she will withdraw every six months. Use Tables 23-2A and 23-2B or a calculator.

SOLUTION:
$36,700 ÷ 15.24696 = $2,407.04 Withdrawn every 6 months

LEARNING OBJECTIVE 4

33.
$4,000 (present value) is deposited today at 9% compounded monthly. Compute the monthly payment that can be withdrawn each month for 2 years and empty the account. Use Tables 23-2A and 23-2B or a calculator.

SOLUTION:
$4,000 ÷ 21.88915 = $182.74 Monthly payments

LEARNING OBJECTIVE 3

34.
An investment guarantees to return a minimum of 9% compounded annually for eleven years. What is the total present value of eleven annual withdrawals of $3,000 each? Use Tables 23-2A and 23-2B or a calculator.

SOLUTION:
$3,000 × 6.80519 = $20,415.57 Present value

35.
Compute the present value of 10 semiannual payments of $500 each, if interest is 10% compounded semiannually. Use Tables 23-2A and 23-2B or a calculator.

SOLUTION:
$500 × 7.72173 = $3,860.87 Present value

LEARNING OBJECTIVE 5

36.

$6,500 (present value) is borrowed today at 12% compounded quarterly. Compute the quarterly payments that are required to exactly pay off the loan in 4 years. Use Tables 23-2A and 23-2B or a calculator.

SOLUTION:
$6,500 ÷ 12.56110 = $517.47 Quarterly payments

37.

Judy Baker borrowed $5,000 and agreed to amortize the loan over 7 months by making monthly payments with interest of 9% compounded monthly on the unpaid balance each month. Compute the size of Judy's monthly payments. Use Tables 23-2A and 23-2B or a calculator.

SOLUTION:
$5,000 ÷ 6.79464 = $735.87 Monthly payments

38.

Marilyn Post borrowed $10,000 from her bank which amortized the loan over 2 1/2 years at a rate of 6% compounded monthly. Find the size of Marilyn's monthly payments. Use Tables 23-2A and 23-2B or a calculator.

SOLUTION:
$10,000 ÷ 27.79405 = $359.79 Monthly payments

39.

Lester Newlander needed to borrow $20,000 for his business. Lester's father agreed to loan him the money and amortize it over 21 months with monthly payments and interest on the unpaid balance of 6% compounded monthly. Compute the size of Lester's payments. Use Tables 23-2A and 23-2B or a calculator.

SOLUTION:
$20,000 ÷ 19.88798 = $1,005.63 Monthly payments

40.

Brendina Gillespie wanted to buy a new car. A bank would loan the sum of $18,000 if Brendina would repay the loan over two years. Compute the size of the monthly payment that Brendina needs to pay in order to amortize the loan with interest of 12% compounded monthly on the unpaid balance. Use Tables 23-2A and 23-2B or a calculator.

SOLUTION:
$18,000 ÷ 21.24339 = $847.32 Monthly payments

LEARNING OBJECTIVE 6

41.

Tanya Cordobes borrowed $9,500 from a bank. The loan was amortized over two years. Tanya made equal monthly payments of $434.00, which included interest on the unpaid balance of 0.75% per month (9% annually). Complete the first two months of the amortization schedule.

Amortization Schedule

	Month	Unpaid Balance	Interest Payment	Total Payment	Principal Payment	New Balance
a.	1	_____	_____	$434.00	_____	_____
b.	2	_____	_____	$434.00	_____	_____

SOLUTION:

	Unpaid Balance	Interest Payment	Principal Payment	New Balance
a.	$9,500.00	$71.25	$362.75	$9,137.25
b.	$9,137.25	$68.53	$365.47	$8,771.78

42.

Jerry Isaacs went to his credit union to inquire about borrowing $8,000. The credit union told him they could amortize an $8,000 loan over two years with 24 payments of $376.59. The interest rate would be 1% per month on the unpaid balance (12% annual rate). Complete the first two months of the amortization schedule.

Amortization Schedule

	Month	Unpaid Balance	Interest Payment	Total Payment	Principal Payment	New Balance
a.	1	_____	_____	$376.59	_____	_____
b.	2	_____	_____	$376.59	_____	_____

SOLUTION:

	Unpaid Balance	Interest Payment	Principal Payment	New Balance
a.	$8,000.00	$80.00	$296.59	$7,703.41
b.	$7,703.41	$77.03	$299.56	$7,403.85

Period (n)	0.50%	0.75%	1.00%	1.25%	1.50%	2.00%	3.00%
1	1.00000	1.00000	1.00000	1.00000	1.00000	1.00000	1.00000
2	2.00500	2.00750	2.01000	2.01250	2.01500	2.02000	2.03000
3	3.01502	3.02256	3.03010	3.03766	3.04522	3.06040	3.09090
4	4.03010	4.04523	4.06040	4.07563	4.09090	4.12161	4.18363
5	5.05025	5.07556	5.10101	5.12657	5.15227	5.20404	5.30914
6	6.07550	6.11363	6.15202	6.19065	6.22955	6.30812	6.46841
7	7.10588	7.15948	7.21354	7.26804	7.32299	7.43428	7.66246
8	8.14141	8.21318	8.28567	8.35889	8.43284	8.58297	8.89234
9	9.18212	9.27478	9.36853	9.46337	9.55933	9.75463	10.15911
10	10.22803	10.34434	10.46221	10.58167	10.70272	10.94972	11.46388
11	11.27917	11.42192	11.56683	11.71394	11.86326	12.16872	12.80780
12	12.33556	12.50759	12.68250	12.86036	13.04121	13.41209	14.19203
13	13.39724	13.60139	13.80933	14.02112	14.23683	14.68033	15.61779
14	14.46423	14.70340	14.94742	15.19638	15.45038	15.97394	17.08632
15	15.53655	15.81368	16.09690	16.38633	16.68214	17.29342	18.59891
16	16.61423	16.93228	17.25786	17.59116	17.93237	18.63929	20.15688
17	17.69730	18.05927	18.43044	18.81105	19.20136	20.01207	21.76159
18	18.78579	19.19472	19.61475	20.04619	20.48938	21.41231	23.41444
19	19.87972	20.33868	20.81090	21.29677	21.79672	22.84056	25.11687
20	20.97912	21.49122	22.01900	22.56298	23.12367	24.29737	26.87037
21	22.08401	22.65240	23.23919	23.84502	24.47052	25.78332	28.67649
22	23.19443	23.82230	24.47159	25.14308	25.83758	27.29898	30.53678
23	24.31040	25.00096	25.71630	26.45737	27.22514	28.84496	32.45288
24	25.43196	26.18847	26.97346	27.78808	28.63352	30.42186	34.42647
25	26.55912	27.38488	28.24320	29.13544	30.06302	32.03030	36.45926
26	27.69191	28.59027	29.52563	30.49963	31.51397	33.67091	38.55304
27	28.83037	29.80470	30.82089	31.88087	32.98668	35.34432	40.70963
28	29.97452	31.02823	32.12910	33.27938	34.48148	37.05121	42.93092
29	31.12439	32.26094	33.45039	34.69538	35.99870	38.79223	45.21885
30	32.28002	33.50290	34.78489	36.12907	37.53868	40.56808	47.57542

Table 23-1A. Future Value of an Ordinary Annuity of $1.00 per Period

Period (n)	4.00%	5.00%	6.00%	8.00%	9.00%	10.00%	12.00%
1	1.00000	1.00000	1.00000	1.00000	1.00000	1.00000	1.00000
2	2.04000	2.05000	2.06000	2.08000	2.09000	2.10000	2.12000
3	3.12160	3.15250	3.18360	3.24640	3.27810	3.31000	3.37440
4	4.24646	4.31013	4.37462	4.50611	4.57313	4.64100	4.77933
5	5.41632	5.52563	5.63709	5.86660	5.98471	6.10510	6.35285
6	6.63298	6.80191	6.97532	7.33593	7.52333	7.71561	8.11519
7	7.89829	8.14201	8.39384	8.92280	9.20043	9.48717	10.08901
8	9.21423	9.54911	9.89747	10.63663	11.02847	11.43589	12.29969
9	10.58280	11.02656	11.49132	12.48756	13.02104	13.57948	14.77566
10	12.00611	12.57789	13.18079	14.48656	15.19293	15.93742	17.54874
11	13.48635	14.20679	14.97164	16.64549	17.56029	18.53117	20.65458
12	15.02581	15.91713	16.86994	18.97713	20.14072	21.38428	24.13313
13	16.62684	17.71298	18.88214	21.49530	22.95338	24.52271	28.02911
14	18.29191	19.59863	21.01507	24.21492	26.01919	27.97498	32.39260
15	20.02359	21.57856	23.27597	27.15211	29.36092	31.77248	37.27971
16	21.82453	23.65749	25.67253	30.32428	33.00340	35.94973	42.75328
17	23.69751	25.84037	28.21288	33.75023	36.97370	40.54470	48.88367
18	25.64541	28.13238	30.90565	37.45024	41.30134	45.59917	55.74971
19	27.67123	30.53900	33.75999	41.44626	46.01846	51.15909	63.43968
20	29.77808	33.06595	36.78559	45.76196	51.16012	57.27500	72.05244
21	31.96920	35.71925	39.99273	50.42292	56.76453	64.00250	81.69874
22	34.24797	38.50521	43.39229	55.45676	62.87334	71.40275	92.50258
23	36.61789	41.43048	46.99583	60.89330	69.53194	79.54302	104.60289
24	39.08260	44.50200	50.81558	66.76476	76.78981	88.49733	118.15524
25	41.64591	47.72710	54.86451	73.10594	84.70090	98.34706	133.33387
26	44.31174	51.11345	59.15638	79.95442	93.32398	109.18177	150.33393
27	47.08421	54.66913	63.70577	87.35077	102.72313	121.09994	169.37401
28	49.96758	58.40258	68.52811	95.33883	112.96822	134.20994	190.69889
29	52.96629	62.32271	73.63980	103.96594	124.13536	148.63093	214.58275
30	56.08494	66.43885	79.05819	113.28321	136.30754	164.49402	241.33268

Table 23-1B. Future Value of an Ordinary Annuity of $1.00 per Period

Period (n)	0.50%	0.75%	1.00%	1.25%	1.50%	2.00%	3.00%
1	0.99502	0.99256	0.99010	0.98765	0.98522	0.98039	0.97087
2	1.98510	1.97772	1.97040	1.96312	1.95588	1.94156	1.91347
3	2.97025	2.95556	2.94099	2.92653	2.91220	2.88388	2.82861
4	3.95050	3.92611	3.90197	3.87806	3.85438	3.80773	3.71710
5	4.92587	4.88944	4.85343	4.81784	4.78264	4.71346	4.57971
6	5.89638	5.84560	5.79548	5.74601	5.69719	5.60143	5.41719
7	6.86207	6.79464	6.72819	6.66273	6.59821	6.47199	6.23028
8	7.82296	7.73661	7.65168	7.56812	7.48593	7.32548	7.01969
9	8.77906	8.67158	8.56602	8.46234	8.36052	8.16224	7.78611
10	9.73041	9.59958	9.47130	9.34553	9.22218	8.98259	8.53020
11	10.67703	10.52067	10.36763	10.21780	10.07112	9.78685	9.25262
12	11.61893	11.43491	11.25508	11.07931	10.90751	10.57534	9.95400
13	12.55615	12.34235	12.13374	11.93018	11.73153	11.34837	10.63496
14	13.48871	13.24302	13.00370	12.77055	12.54338	12.10625	11.29607
15	14.41662	14.13699	13.86505	13.60055	13.34323	12.84926	11.93794
16	15.33993	15.02431	14.71787	14.42029	14.13126	13.57771	12.56110
17	16.25863	15.90502	15.56225	15.22992	14.90765	14.29187	13.16612
18	17.17277	16.77918	16.39827	16.02955	15.67256	14.99203	13.75351
19	18.08236	17.64683	17.22601	16.81931	16.42617	15.67846	14.32380
20	18.98742	18.50802	18.04555	17.59932	17.16864	16.35143	14.87747
21	19.88798	19.36280	18.85698	18.36969	17.90014	17.01121	15.41502
22	20.78406	20.21121	19.66038	19.13056	18.62082	17.65805	15.93692
23	21.67568	21.05331	20.45582	19.88204	19.33086	18.29220	16.44361
24	22.56287	21.88915	21.24339	20.62423	20.03041	18.91393	16.93554
25	23.44564	22.71876	22.02316	21.35727	20.71961	19.52346	17.41315
26	24.32402	23.54219	22.79520	22.08125	21.39863	20.12104	17.87684
27	25.19803	24.35949	23.55961	22.79630	22.06762	20.70690	18.32703
28	26.06769	25.17071	24.31644	23.50252	22.72672	21.28127	18.76411
29	26.93302	25.97589	25.06579	24.20002	23.37608	21.84438	19.18845
30	27.79405	26.77508	25.80771	24.88891	24.01584	22.39646	19.60044

Table 23-2A. Present Value of an Ordinary Annuity of $1.00 per Period

Period (n)	4.00%	5.00%	6.00%	8.00%	9.00%	10.00%	12.00%
1	0.96154	0.95238	0.94340	0.92593	0.91743	0.90909	0.89286
2	1.88609	1.85941	1.83339	1.78326	1.75911	1.73554	1.69005
3	2.77509	2.72325	2.67301	2.57710	2.53129	2.48685	2.40183
4	3.62990	3.54595	3.46511	3.31213	3.23972	3.16987	3.03735
5	4.45182	4.32948	4.21236	3.99271	3.88965	3.79079	3.60478
6	5.24214	5.07569	4.91732	4.62288	4.48592	4.35526	4.11141
7	6.00205	5.78637	5.58238	5.20637	5.03295	4.86842	4.56376
8	6.73274	6.46321	6.20979	5.74664	5.53482	5.33493	4.96764
9	7.43533	7.10782	6.80169	6.24689	5.99525	5.75902	5.32825
10	8.11090	7.72173	7.36009	6.71008	6.41766	6.14457	5.65022
11	8.76048	8.30641	7.88687	7.13896	6.80519	6.49506	5.93770
12	9.38507	8.86325	8.38384	7.53608	7.16073	6.81369	6.19437
13	9.98565	9.39357	8.85268	7.90378	7.48690	7.10336	6.42355
14	10.56312	9.89864	9.29498	8.24424	7.78615	7.36669	6.62817
15	11.11839	10.37966	9.71225	8.55948	8.06069	7.60608	6.81086
16	11.65230	10.83777	10.10590	8.85137	8.31256	7.82371	6.97399
17	12.16567	11.27407	10.47726	9.12164	8.54363	8.02155	7.11963
18	12.65930	11.68959	10.82760	9.37189	8.75563	8.20141	7.24967
19	13.13394	12.08532	11.15812	9.60360	8.95011	8.36492	7.36578
20	13.59033	12.46221	11.46992	9.81815	9.12855	8.51356	7.46944
21	14.02916	12.82115	11.76408	10.01680	9.29224	8.64869	7.56200
22	14.45112	13.16300	12.04158	10.20074	9.44243	8.77154	7.64465
23	14.85684	13.48857	12.30338	10.37106	9.58021	8.88322	7.71843
24	15.24696	13.79864	12.55036	10.52876	9.70661	8.98474	7.78432
25	15.62208	14.09394	12.78336	10.67478	9.82258	9.07704	7.84314
26	15.98277	14.37519	13.00317	10.80998	9.92897	9.16095	7.89566
27	16.32959	14.64303	13.21053	10.93516	10.02658	9.23722	7.94255
28	16.66306	14.89813	13.40616	11.05108	10.11613	9.30657	7.98442
29	16.98371	15.14107	13.59072	11.15841	10.19828	9.36961	8.02181
30	17.29203	15.37245	13.76483	11.25778	10.27365	9.42691	8.05518

Table 23-2B. Present Value of an Ordinary Annuity of $1.00 per Period

Chapter 24 BUSINESS STATISTICS

PROBLEMS

LEARNING OBJECTIVES 1, 2, 3

1.
Compute the mean, median, and mode of the numbers in data set below. Round answers to the nearest 1/10.

58 28 15 30 48 32 17 32 29

a.   Mean: _____
b.   Median: _____
c.   Mode: _____

SOLUTION:

15 + 17 + 28 + 29 + 30 + 32 + 32 + 48 + 58 = 289

a. $289 \div 9 = 32.1$
b.   30
c.   32

2.
Compute the mean, median, and mode of the numbers in data set below. Round answers to the nearest 1/10.

Set B: 30 25 16 59 34 44 16 25 59 29 25

a.   Mean: _____
b.   Median: _____
c.   Mode: _____

SOLUTION:

16 + 16 + 25 + 25 + 25 + 29 + 30 + 34 + 44 + 59 + 59 = 362

a. $362 \div 11 = 32.9$
b.   29
c.   25

3.
Compute the mean, median, and mode of the numbers in data set below. Round answers to the nearest 1/10.

17 28 15 30 48 58 34 17 32 29

a.   Mean: _____
b.   Median: _____
c.   Mode: _____

SOLUTION:
        15 + 17 + 17 + 28 + 29 + 30 + 32 + 34 + 48 + 58 = 308
a.   308 ÷ 10 = 30.8
b.   (29 + 30) ÷ 2 = 29.5
c.   17

4.
Compute the mean, median, and mode of the numbers in data set below. Round answers to the nearest 1/10.

Set B: 51 25 59 27 16 59 30 25 16 59 34 44

a.   Mean: _____
b.   Median: _____
c.   Mode: _____

SOLUTION:
        16 + 16 + 25 + 25 + 27 + 30 + 34 + 44 + 51 + 59 + 59 + 59 = 445
a.   445 ÷ 12 = 37.1
b.   (30 + 34) ÷ 2 = 32
c.   59

5.
Telemarketers were hired by the city newspaper to telephone city residents and offer them an introductory subscription to the newspaper. The telemarketers were supposed to call between 6:00 p.m. and 7:00 p.m. in the evening, Monday through Thursday, for three weeks. Shown below are the numbers of subscribers contacted between 6:00 and 7:00 p.m. Compute the mean, median, and mode of those subscribers contacted. Round answers to the nearest 1/10.

	M	T	W	Th	M	T	W	Th	M	T	W	Th
Subscribers contacted:	12	23	12	17	14	21	29	21	18	21	25	13

a.   Mean: _____
b.   Median: _____
c.   Mode: _____

SOLUTION:

   12 + 12 + 13 + 14 + 17 + 18 + 21 + 21 + 21 + 23 + 25 + 29 = 226

a. $226 \div 12 = 18.8$

b. $(18 + 21) \div 2 = 19.5$

c.   21

6.

Telemarketers were hired by the city newspaper to telephone city residents and offer them an introductory subscription to the newspaper. The telemarketers were supposed to call between 7:00 p.m. and 8:00 p.m. in the evening, Monday through Thursday, for three weeks. Shown below are the numbers of subscribers contacted between 7:00 and 8:00 p.m. Compute the mean, median, and mode of those subscribers contacted. Round answers to the nearest 1/10.

	M	T	W	Th	M	T	W	Th	M	T	W	Th
Subscribers contacted:	29	31	21	23	38	28	29	25	33	27	35	29

a.   Mean: _____

b.   Median: _____

c.   Mode: _____

SOLUTION:

   21 + 23 + 25 + 27 + 28 + 29 + 29 + 29 + 31 + 33 + 35 + 38 = 348

a. $348 \div 12 = 29$

b. $(29 + 29) \div 2 = 29$

c.   29

7.

Telemarketers were hired by the city newspaper to telephone city residents and offer them an introductory subscription to the newspaper. The telemarketers were supposed to call between 8:00 p.m. and 9:00 p.m. in the evening, Monday through Thursday, for three weeks. Shown below are the numbers of subscribers contacted between 8:00 and 9:00 p.m. Compute the mean, median, and mode of those subscribers contacted. Round answers to the nearest 1/10.

	M	T	W	Th	M	T	W	Th	M	T	W	Th
Subscribers contacted:	24	35	26	33	30	34	28	24	21	31	24	30

a.   Mean: _____

b.   Median: _____

c.   Mode: _____

SOLUTION:

   21 + 24 + 24 + 24 + 26 + 28 + 30 + 30 + 31 + 33 + 34 + 35 = 340

a. $340 \div 12 = 28.3$

b. $(28 + 30) \div 2 = 29$

c.   24

LEARNING OBJECTIVE 4

8.

Construct a frequency distribution for the set of data below. Use four classes, beginning with "10 up to 18," then "18 up to 24," then "24 up to 30," and finally "30 up to 36."

Data Set

16	35	28	13	32
28	22	13	21	20
33	17	33	15	23
35	31	29	14	25
25	32	31	27	32

Frequency Distribution

	Class Interval	Tally	Frequency
a.	_____	_____	_____
b.	_____	_____	_____
c.	_____	_____	_____
d.	_____	_____	------------
	Total		_____

e.   Compute the mean of the data set. Check your work by adding both the row totals and the column totals.

SOLUTION:

Frequency Distribution

	Class Interval	Tally	Frequency
a.	10 up to 18	ℍℍ I	6
b.	18 up to 24	IIII	4
c.	24 up to 30	ℍℍ I	6
d.	30 up to 36	ℍℍ IIII	9
	Total		25
e.	25.2 (630 ÷ 25) = mean		

9.

Construct a frequency distribution for the set of data below. Use four classes, beginning with "20 up to 28," then "28 up to 36," then "36 up to 44," and finally "44 up to 52."

Data Set

39	27	21	51	38	22
27	48	31	27	42	35
43	42	45	50	35	27
32	37	22	38	23	33

Frequency Distribution

	Class Interval	Tally	Frequency
a.	_____	_____	_____
b.	_____	_____	_____

c. _____   _____   _____
d. _____   _____   _____

Total   ------------------
                                                       _____

e.   Compute the mean of the data set. Check your work by adding both the row totals and the column totals.

SOLUTION:

                                  Frequency Distribution
        Class Interval   Tally   Frequency
a.   20 up to 28   ⊬⊬ ||| 8
b.   28 up to 36   ⊬⊬   5
c.   36 up to 44   ⊬⊬ || 7
d.   44 up to 52 |||| 4
        Total   24
e.   34.8 (835 ÷ 24) mean

10.

Construct a frequency distribution for the set of data below. Use four classes, beginning with "10 up to 20," then "20 up to 30," then "30 up to 40," and finally "40 up to 50."

                                      Data Set
        19   25   45   48   12
        28   42   13   41   29
        33   46   43   18   14
        35   11   29   48   25
        24   32   38   27   32
        19   16   33   14   22

                                  Frequency Distribution
        Class Interval   Tally   Frequency
a.   _____   _____   _____
b.   _____   _____   _____
c.   _____   _____   _____
d.   _____   _____   _____

        Total   ------------------
                                                             _____

e.   Compute the mean of the data set. Check your work by adding both the row totals and the column totals.

SOLUTION:

                                  Frequency Distribution
        Class Interval   Tally   Frequency
a.   10 up to 20   ⊬⊬ |||| 9
b.   20 up to 30   ⊬⊬ ||| 8
c.   30 up to 40   ⊬⊬ | 6
d.   40 up to 50   ⊬⊬ || 7
        Total   30
e.   28.7 (861 ÷ 30) mean

11.

Construct a frequency distribution for the set of data below. Use five classes, beginning with "20 up to 30," then "30 up to 40," then "40 up to 50," then "50 up to 60," and finally "60 up to 70.

Data Set

39	28	21	38	38	62
67	48	22	27	58	51
44	42	64	62	35	26
32	36	52	28	53	34
26	42	26	27	41	49
22	45	37	55	33	59

Frequency Distribution

	Class Interval	Tally	Frequency
a.	_____	_____	_____
b.	_____	_____	_____
c.	_____	_____	_____
d.	_____	_____	_____
e.	_____	_____	_____
	Total		_____

f.   Compute the mean of the data set. Check your work by adding both the row totals and the column totals.

SOLUTION:

Frequency Distribution

	Class Interval	Tally	Frequency
a.	20 up to 30	₩ ₩	10
b.	30 up to 40	₩ IIII	9
c.	40 up to 50	₩ II	7
d.	50 up to 60	₩ I	6
e.	60 up to 70	IIII	4
	Total		36

e.   40.8 (1,469 ÷ 36) mean

12.

Angela's Meat & Poultry, a chain of specialty meat markets, is taking a survey to determine whether customers' demands are changing with respect to preferences for beef, pork, and chicken. They record the data shown below from recent sales of beef. Construct a frequency distribution for beef sales. Use four classes, beginning with "0 up to 1," then "1 up to 2," and so on.

Pounds of Beef Sold

3.3	0.6	1.5	1.8	2.3	1.7	1.2	2.8	1.7
1.9	3.2	2.6	0.9	1.7	1.4	3.9	1.9	0.8
1.2	2.1	0.8	1.8	3.6	2.2	2.8	3.7	0.6
0.6	3.2	3.5	0.9	1.5	2.8	3.9	0.9	1.4

Frequency Distribution

	Class Interval	Tally	Frequency
a.	_____	_____	_____
b.	_____	_____	_____
c.	_____	_____	_____
d.	_____	_____	_____
Total			_____

e.   Compute the mean pounds of beef sold. Check your work by adding both the row totals and column totals.

SOLUTION:

Frequency Distribution

	Class Interval	Tally	Frequency
a.	0 up to 1	⦀⦀ ⦀⦀⦀	8
b.	1 up to 2	⦀⦀ ⦀⦀ ⦀⦀⦀	13
c.	2 up to 3	⦀⦀ ⦀⦀	7
d.	3 up to 4	⦀⦀ ⦀⦀⦀	8
	Total		36

e.   2.02 lb. ($72.7 \div 36$) mean

13.

Angela's Meat & Poultry, a chain of specialty meat markets, is taking a survey to determine whether customers' demands are changing with respect to preferences for beef, pork, and chicken. They record the data shown below from recent sales of pork. Construct a frequency distribution for pork sales. Use four classes, beginning with "0 up to 1," then "1 up to 2," and so on.

Pounds of Pork Sold

3.2	1.5	2.2	0.6	2.1	0.5	2.3
1.1	0.4	3.1	1.1	3.2	1.2	2.1
3.1	3.4	2.1	3.3	2.5	2.4	3.1
0.4	2.1	1.2	3.1	0.5	1.3	0.6

Frequency Distribution for Pork Sales

	Class Interval	Tally	Frequency
a.	_____	_____	_____
b.	_____	_____	_____
c.	_____	_____	_____
d.	_____	_____	_____
Total			_____

e.   Compute the mean pounds of pork sold. Check your work by adding both the row totals and the column totals.

SOLUTION:

Frequency Distribution for Pork Sales

	Class Interval	Tally	Frequency
a.	0 up to 1	ⅢⅡ I	6
b.	1 up to 2	ⅢⅡ I	6
c.	2 up to 3	ⅢⅡ III	8
d.	3 up to 4	ⅢⅡ III	8
	Total		28

e.   1.92 lbs. (53.7 ÷ 28) mean

14.

Angela's Meat & Poultry, a chain of specialty meat markets, is taking a survey to determine whether customers' demands are changing with respect to preferences for beef, pork and chicken. They record the data shown below from recent sales of chicken. Construct a frequency distribution for chicken sales. Use four classes, beginning with "0 up to 1," then "1 up to 2," and so on.

Pounds of Chicken Sold

0.4	1.8	2.5	0.6	2.1	0.8	2.6	3.3
2.8	0.6	3.1	1.4	3.3	1.5	2.1	1.3
3.5	3.7	2.2	3.6	2.8	2.7	2.4	3.4
1.1	2.3	1.4	3.2	0.7	1.8	0.8	0.7

Frequency Distribution for Chicken Sales

	Class Interval	Tally	Frequency
a.	_____	_____	_____
b.	_____	_____	_____
c.	_____	_____	_____
d.	_____	_____	_____
	Total		_____

e.   Compute the mean pounds of chicken sold. Check your work by adding both the row totals and the column totals.

SOLUTION:

Frequency Distribution for Chicken Sales

	Class Interval	Tally	Frequency
a.	0 up to 1	ⅢⅡ II	7
b.	1 up to 2	ⅢⅡ II	7
c.	2 up to 3	ⅢⅡ ⅢⅡ	10
d.	3 up to 4	ⅢⅡ III	8
	Total		32

e.   2.08 lbs. (66.5 ÷ 32) mean

LEARNING OBJECTIVE 5

15.

Use the frequency distribution shown below and Figure 24-1 to construct a histogram which displays the data.

Frequency Distribution

	Class Interval	Frequency
a.	100 up to 200	12
b.	200 up to 300	18
c.	300 up to 400	12
d.	400 up to 500	8
	Total	50

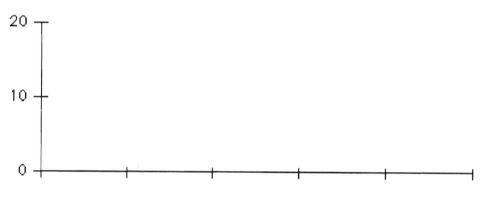

FIGURE 24-1

SOLUTION:

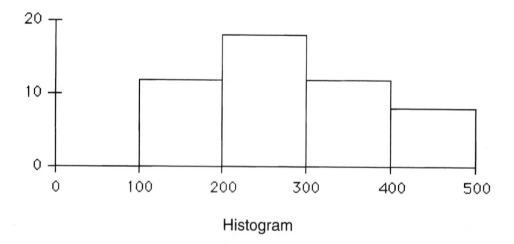

Histogram

16.

Enrico Vaselli owns Enrico's European Shoe Store, a chain of small stores that sell primarily Italian and French shoes. The operations manager of one store made a frequency distribution of the amounts of 75 randomly selected previous sales, as shown in the following table. Use Figure 24-2 to construct a histogram to display the data.

Enrico's European Shoe Store
Shoe Sales

	Class Interval	Frequency
a.	$ 25 up to $ 75	15
b.	$ 75 up to $125	30
c.	$125 up to $175	20
d.	$175 up to $225	10
	Total	75

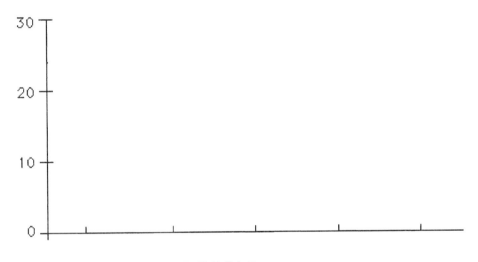

FIGURE 24-2

SOLUTION:

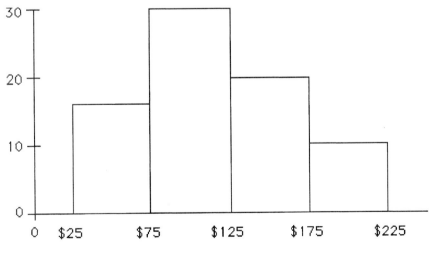

Enrico's European Shoe Store
Shoe Sales

17.
Use Figure 24-3 to construct a histogram for the frequency distribution in the table below. Write labels where necessary.

Frequency Distribution

Class Interval	Frequency
0 up to 100	9
100 up to 200	13
200 up to 300	15
300 up to 400	12
400 up to 500	<u>11</u>
Total	60

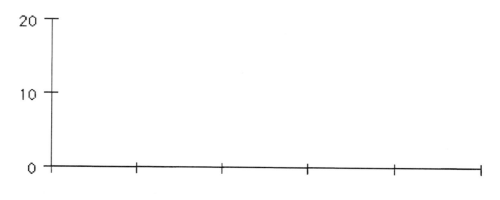

FIGURE 24-3

SOLUTION:

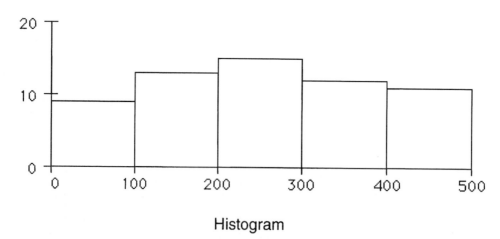

Histogram

18.

In June, Karen Henderson, an accountant, examined her records to learn the amount she had billed each of the first 90 clients who had contacted her at the beginning of the year for income tax advice. The results of the survey appear in the following table. Use Figure 24-4 to construct a histogram to illustrate this data. Write labels where necessary.

Karen Henderson, Accountant
Billing Amounts

Class Interval	Frequency
$ 0 up to $ 50	24
$ 50 up to $100	12
$100 up to $150	14
$150 up to $200	21
$200 up to $250	19
Total	90

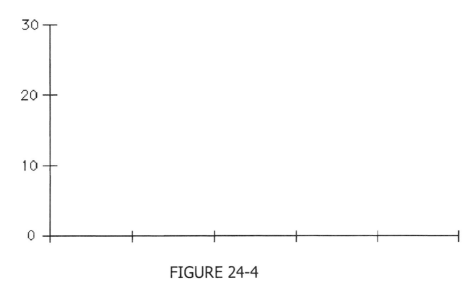

FIGURE 24-4

SOLUTION:

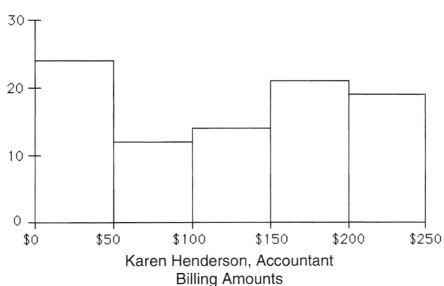

Karen Henderson, Accountant
Billing Amounts

LEARNING OBJECTIVES 6, 7

19.

The table below displays the quarterly revenue of Wallace Printing for this year, arranged by quarter. The numbers are all in thousands of dollars. Use Figure 24-5 to construct a bar graph and a line graph for the data. Write labels where necessary.

Wallace Printing

Quarter	Revenue (in thousands of $) This Year
Jan–Mar	$275
Apr–June	$225
Jul–Sept	$300
Oct–Dec	$275

a.   Bar graph

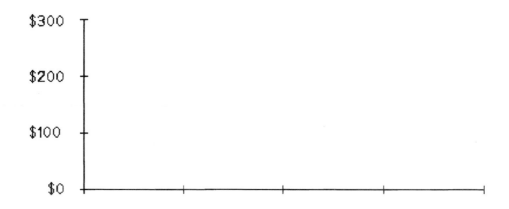

b.   Line graph

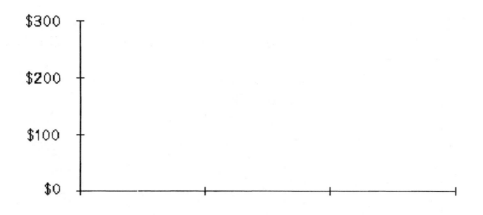

FIGURE 24-5

SOLUTION:

a.   Bar graph

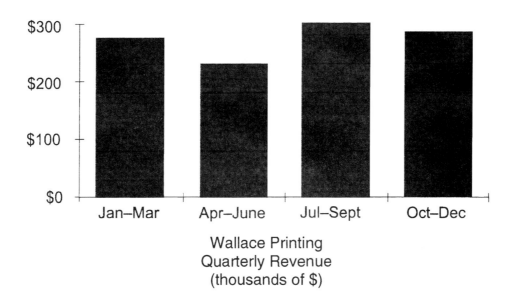

b.   Line graph

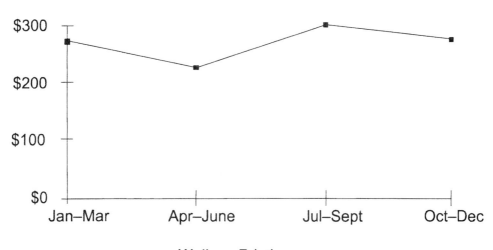

LEARNING OBJECTIVE 6

20.

The table below displays the quarterly revenues of Wallace Printing for this year and last year. The numbers are all in thousands of dollars. Use Figure 24-6 to construct a comparative bar graph for the data. Write labels where necessary. Shade the bars for each year differently.

Wallace Printing

	Revenue (in thousands of $)	
Quarter	This Year	Last Year
Jan–Mar	$275	$200
Apr–June	$225	$175
Jul–Sept	$300	$225
Oct–Dec	$275	$250

Comparative bar graph

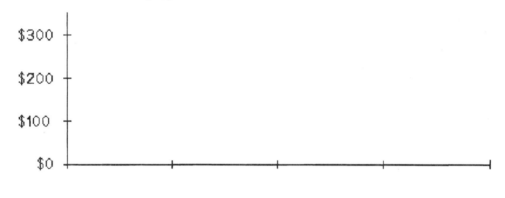

FIGURE 24-6

SOLUTION:

Comparative bar graph

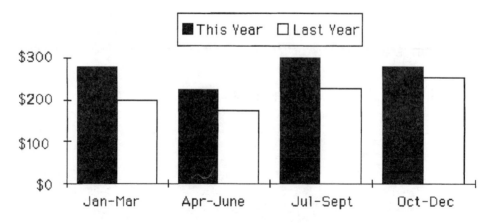

Wallace Printing
Quarterly Revenue (in thousands of dollars)
This Year and Last Year

LEARNING OBJECTIVES 6, 7

21.
Last year David Lum opened a new pharmacy. He already had one store, located near the city hospital. When a new outpatient surgery clinic began operation, he opened his second pharmacy nearby. In June of this year, Lum wanted a business loan. The bank asked him to report the monthly revenues of each store for the first five months of the year. Lum displayed the data as follows.

David Lum Pharmacies
Monthly Revenue: January–May
(thousands of dollars)

	Jan	Feb	Mar	Apr	May
Hospital Store	$62	$64	$61	$63	$64
Clinic Store	$30	$32	$36	$34	$40

a.   Use Figure 24-7a to construct a bar graph to illustrate the monthly revenues of the clinic store. Write labels where necessary.

b   Use Figure 24-7b to construct a line graph to illustrate the monthly revenues of the clinic store. Write labels where necessary.

c.   Use Figure 24-7c to construct a comparative bar graph to illustrate the monthly revenues of both pharmacies. Write labels where necessary. Shade the bars for each store differently.

a.   Bar graph

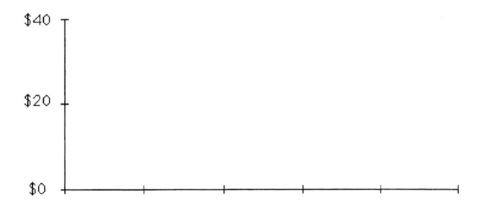

FIGURE 24-7a

b.   Line graph

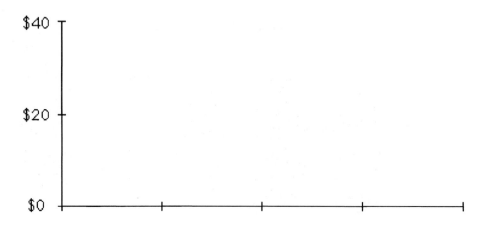

FIGURE 24-7b

c.   Comparative bar graph

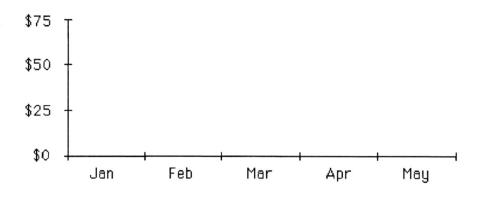

FIGURE 24-7c

SOLUTION:

a.   Bar graph

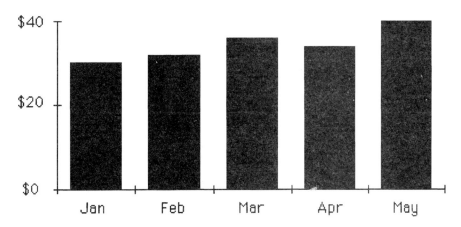

David Lum Pharmacies
Outpatient Clinic Store
Monthly Revenue: January–May
(thousands of $)

b.   Line graph

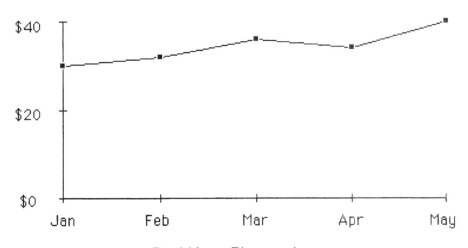

David Lum Pharmacies
Outpatient Clinic Store
Monthly Revenue: January–May
(thousands of $)

c.   Comparative bar graph

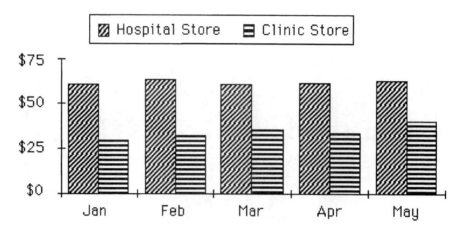

David Lum Pharmacies
Hospital and Clinic Store
Monthly Revenue: January–May

LEARNING OBJECTIVE 6

22.

David Lum operates two pharmacies, one near the hospital and a second near the outpatient surgery clinic. Lum categorizes sales revenues as prescription medicine, over-the-counter medicine, or non-medical sales. The monthly revenues from each store for the first five months of the year are shown in the table below. Use Figures 24-8a and 24-8b to construct two component bar graphs, one for each store, to illustrate the data. Write labels where necessary.

David Lum Pharmacies
Hospital Store
Monthly Revenue: January–May
(thousands of dollars)

	Jan	Feb	Mar	Apr	May
Prescription	$38	$43	$40	$36	$40
Over-the-Counter	15	16	12	15	16
Non-Medical	9	5	9	12	8
Totals	$62	$64	$61	$63	$64

a.   Component bar graph, Hospital Store

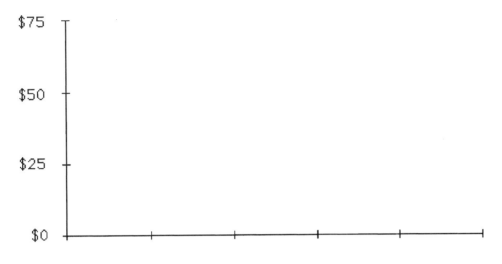

FIGURE 24-8a

David Lum Pharmacies
Outpatient Clinic Store
Monthly Revenue: January–May
(thousands of dollars)

	Jan	Feb	Mar	Apr	May
Prescription	$17	$15	$19	$15	$20
Over-the-Counter	9	12	13	12	15
Non-Medical	4	5	4	7	5
Totals	$30	$32	$36	$34	$40

b.   Component bar graph, Outpatient Clinic Store

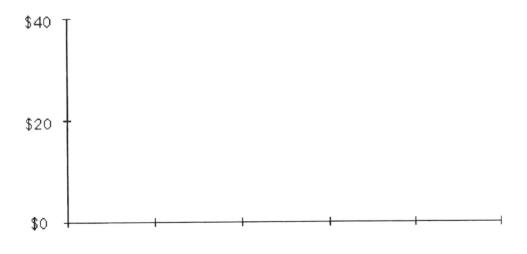

FIGURE 24-8b

SOLUTION:

a.   Component bar graph, Hospital Store

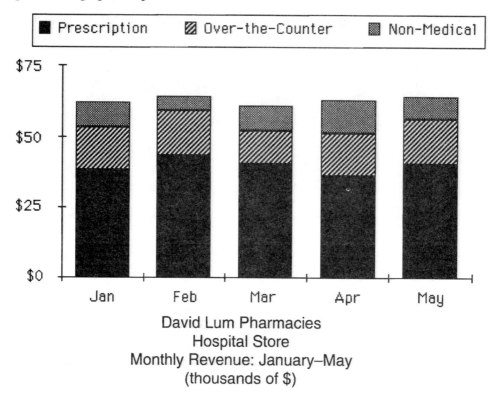

b.   Component bar graph, Outpatient Clinic Store

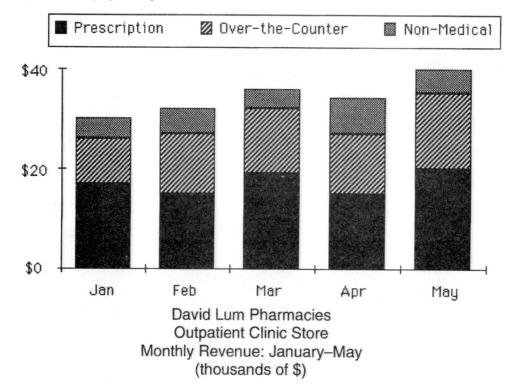

LEARNING OBJECTIVE 8

23.

David Lum operates two pharmacies, one near the hospital and a second near the outpatient surgery clinic. Lum categorizes sales revenues as prescription medicine, over-the-counter medicine, or non-medical sales. The revenues from each store for the month of May are shown below. Using Figures 24-9a and 24-9b construct one circle graph for each store to illustrate the data. Write labels where necessary.

David Lum Pharmacies
Hospital and Clinic Stores
May Revenue
(thousands of dollars)

	Hospital	Clinic
Prescription	$40	$20
Over-the-Counter	16	15
Non-Medical	8	5
Totals	$64	$40

a.   Circle graph, Hospital Store

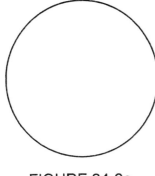

FIGURE 24-9a

	Revenue	Percent	Fraction
Prescription	$40	_____	_____
Over-the-Counter	16	_____	_____
Non-Medical	8	_____	_____
	$64	_____	_____

b.   Circle graph, Clinic Store

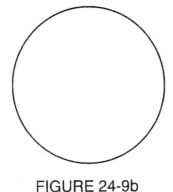

FIGURE 24-9b

	Revenue	Percent	Fraction
Prescription	$20	_____	_____
Over-the-Counter	15	_____	_____
Non-Medical	5	_____	_____
	$40	_____	_____

SOLUTION:

a.   Hospital Store

	Revenue	Percent	Fraction	
Prescription	$40	62.5%	5/8	$40 ÷ $64 = 0.625
Over-the-Counter	16	25.0	1/4	$16 ÷ $64 = 0.25
Non-Medical	8	12.5	1/8	$ 8 ÷ $64 = 0.125
	$64	100.0%	8/8 = 1	

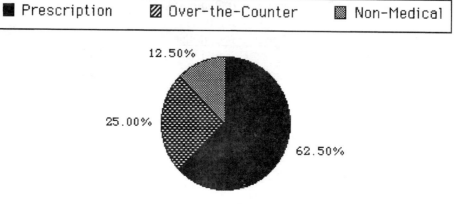

David Lum Pharmacies
May Revenue, Hospital Store

b.   Clinic Store

	Revenue	Percent	Fraction	
Prescription	$20	50.0%	1/2	$20 ÷ $40 = 0.50
Over-the-Counter	15	37.5	3/8	$15 ÷ $40 = 0.375
Non-Medical	5	12.5	1/8	$ 5 ÷ $40 = 0.125
	$40	100.0%	8/8 = 1	

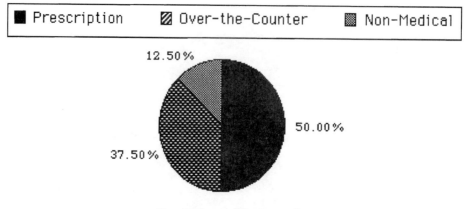

David Lum Pharmacies
May Revenue, Clinic Store